Farming Soul

Farming Soul

a tale of initiation

Patricia Damery

LEAPING GOAT PRESS

Published simultaneously in Canada, the United Kingdom, and the United States of America. For information on obtaining permission for use of material from this work, please submit a written request to:

Leaping Goat Press
3185 Dry Creek Road
Napa, CA 94558
www.leapinggoatpress.com

To protect the privacy of certain persons, the author has fictionalized some names and identifying information. Although obscuring the identity of these individuals, the resulting accounts remain real-life experiences. Any similarity between these personal descriptions and an actual individual is purely coincidental.

Many thanks to all sources that have directly or indirectly provided permission to quote their works, including:

The poem "Song of the Sky Loom " is from "Songs of the Tewa," edited by Herbert Joseph Spinden appears courtesy of Sunstone Press, Box 2321, Santa Fe, NM 87504 www. sunstone-press.com

Chapter 17 first appeared as " Distillation, Donald, and Kant," in Jung Journal:Culture and Psyche, Vol.3, # 3, 2009., UC Press Journals

From The Kabir Book by Robert Bly, Copyright© 1971, 1977 by Robert Bly, © by the Seventies Press, Reprinted by permission of Beacon Press, Boston.

For my seven generations,
And especially,
Jesse and Lisa, Casey and Melissa,
Wesley and Sabien,
And the Ones to come.

Contents

Foreword

by Robert Sardello

Goodness! What a writing! A deliciously compelling story, filled with profound insight concerning the most invisible, and therefore most significant dimensions, of life. I hope to entice you to read this book in a contemplative spirit, for it can reorient your life.

The many stories within the larger story are filled with interesting, provocative, complex characters. I am most touched with the presence of one figure in particular, Natalio. See who touches you in this way, and he or she will be your soul guide through the beautifying imagination of Patricia Damery.

You will discover the intertwining tales of a struggle to become "soul-certified" through analytical psychology training. There is also the joyfully painful story of combining the soul path of depth psychology with the wisdom of American Indian shamanism. This struggle is intertwined with the cultivation of the land, and Damery's deepening sense of being one with nature. She is in training to be both a seer and healer, and a practitioner of Biodynamic agriculture.

An invisible unfolding occurs that touches the deepest desire of the soul. The true author of *Farming Soul* is Soul herself. What differentiates this book from being an autobiography is the invitation to enter a unique form of initiation, one that seems so suitable to this age, this time, our given circumstances, now.

This story is really a myth, a myth of the future. The future is not out in front of us; it is the becoming nature of our being and that of earth as one. This important current weaves through this writing: none of what happens at the body, soul, or spirit dimension separates from earth. This future-as-present myth senses, perceives, feels, and orients always to earth as living Soul Presence. Damery works at dispelling the

I

collective error of taking earth as separate from us, separate from our deep desires and longings as human beings.

The cultural error of separation installs fear and survival at the basis of our existence, an error leading to the world-destructiveness now imploding us. It has not yet dawned that this moment of earth and world crisis is completely founded on exclusion of the fullness of reality. Such an error of sensing—sensing according to belief rather than according to body-immediacy—has resulted in the terrible split between human and earth. This division would be only unfortunate, were it not for the fact that every human, separated in this way, unknowingly colludes part and parcel with the scheme of obliterating the invisible and the world-creative, or relegates it to the regions of spiritual paths or now calcified religions and theologies. Damery is one of the new initiates who locates the all within the all, which is utterly necessary if we are to avoid the all against the all.

A "listening" reading helps anyone discover the ever-present mythic sense of the future and how this differs from established Spiritual Ways. The Way here is not expressed in any of these Ways, but her own unique improvisation. Damery dips to the bottom of many of these established Ways—depth psychology and analysis, shamanism, anthroposophy, and esoteric healing. She did not find any of them lacking, but something quite mysterious must have been burning within Damery, an inner voice, an inner urgency to the discovery that our soul is also World Soul.

The most marvelous dimension of the book centers on the fact that it is not a philosophical thesis; what burns within Damery does not stem from theory. The very nature of story is that it is something other than mental understanding. We are given a true story. True here does not refer to the fact that actual life events are told as story, but as the way things really are, something that can never be apprehended through the intellect alone. Story allows the very real and objective realm of feeling to speak. And feeling knows things that intellect alone can never know, just as we know things through listening to a symphony or contemplating a work of art that can never be known through intellect alone. The added dimension of this story is that it is also highly intelligent, but not abstractly so. Feeling-intelligence originates simultaneously within the real and actual organ of the heart, the center of our being that is simultaneously flesh, soul, and spirit. Feeling thus does not mean that Damery speaks her feelings. It is far more than that, for the personal here is impersonal, but more intimate than anything merely personal could ever be.

The intelligence of the heart is the domain of intuitive awareness; awareness touching itself and that is thus aware of itself without separating from the touched. Take feeling in its most immediately given sense, as knowing through touching, as when one feels wood and knows it is different from the feel of iron. Feeling is the domain of sensing that is self-aware. Heart that is aware, sensing that is aware, precedes abstract thinking and assures that thinking so grounded will be truth— not theory, not abstraction, not speculation, not inquiry, but truth. I feel, therefore earth lives!

The word "consciousness" arrives historically very late, first used to indicate a separation between the mind and what the mind apprehends. Consciousness, thus, is always a consciousness of something. With feeling, no such separation exists. Feeling knows through inner touch, even if what feeling touches is miles away. In earlier times, this way of knowing was called common sense, meaning all of the senses working in common, together, to feel the fullness of reality. The intelligence of the heart magnificently rings and radiates from this story with a truth so true that it is possible to re-found one's life on it. This writing is one of the new world-initiation forms. Thus, we feel every word as living, and we feel the sacrifice, a giving for the benefit of the all.

There is much more in this feast of a book. The author also discovers there is really no split between spirit and soul, that the essence of the human being, what is inherently given, is the fact our deepest desires are completely united with our highest longings.

What so clearly burns within the author is the initiatory gift to continue to work against the given. It took her many years to differentiate. Those years are significant, and not just for her, for what she has accomplished. That living imagination has now entered the world. It is present within the world as a living and effective soul/spirit current.

III

What she went through, with extraordinary effort and suffering, and what then awakens, can now be accomplished by others. She is a maker of spiritual culture, the now arriving, beckoning form of future earth life.

Robert Sardello, Ph.D. is co-director of the School of Spiritual Psychology, Editor of Goldenstone Press, and author of seven books, including Silence: The Mystery of Wholeness.

Preface to the Second Edition

Years ago, while in our analytic training to be Jungian analysts, my candidate group met for a series of seminars in the home of one of our founding members, Dr. Joseph Henderson. Dr. Henderson was in his early 90s at the time and lived on a hilltop in a south-facing, glass-walled home. All morning we gathered in his living room as he talked about individuation and initiation, archetypal processes he spent a lifetime exploring. At noon he sent out for sandwiches, which we ate while continuing our discussion, gazing into the garden and the forested hills beyond.

I do not remember details of those meetings, a phenomenon that I often notice with some of the pivotal teachers in my life. I do remember him saying: *individuation begins when personal analysis ends.* At the time I was shocked. Like most candidates, I was immersed in my analysis, as all candidates were expected to be. I wanted to disagree with him; were we not all individuating (i.e., on the long path of "becoming ourselves")? Isn't that what analysis is, a tool to assist in individuation?

But as the years have gone by, after the deaths of two spiritual teachers and the failure of my analytic work to contain what my soul had in mind for me, all at the heart of my own tale of initiation, I see what Dr. Henderson meant. You may be blessed with good teachers and mentors along the way; you may have had the benefit of a psychotherapy that heals wounds, making it possible to be more of yourself, but in the end, individuation means you are on your own. Analysts and teachers may give direction and help clear your path, but that is all they can do. Individuation is a lonely affair.

Anthropologically, initiation is an archetypal ritual (a rite of passage pattern common to many cultures) in which a child or adolescent "dies" to the old, familiar world of childhood and the mother, and after a period of liminality, a state described by anthropologist Victor Turner as a time of disorienting withdrawal from society and ordeal,[1] is reborn into adulthood and his or her place in the community. Over his many years, Dr. Henderson studied symbols of anthropological initiation and the dreams of his patients, recognizing common factors in the psyche under the archetype. He talked about initiation in broader terms as *initiation in service of individuation*.[2]

Alchemy uses other words to describe this process in relationship to *formation of the philosophical stone*, or the subtle body (which I will discuss more thoroughly in *Farming Soul*, and particularly in Chapters Seven and Eleven). This process most often occurs in the second part of life, which Jung designated beginning after one is established in the world, at about ages 35-40. The ego, too hardened from immersion in the material world and its demands, undergoes various operations, including dissolving (*solutio*) or burning (*calcinatio*), both processes of dismembering of the old way. What used to work no longer does. Thus begins the uncomfortable ordeal of death and decay of old ego attitudes not related to the Self (*mortifactio/nigredo*.) The outcome is not known, nor is it guaranteed! After all, initiations *can* be failed, as I came to know, which may also be part of the ordeal.

Only in enduring is there any hope of rebirth into a larger experience of oneself. This is an individuating experience. Throughout this *tale* I will elaborate on this as I experienced it. An aspect of this essentially religious experience, so often neglected in our present age, is that of gaining a larger perspective in which we re-member our spiritual selves: that we are Spirit in Matter. This re-membering necessitates developing new "organs" or transformers of psychic energy (chakras), part of the formation of the subtle body. Many would say this *is* the development of consciousness.

Do we grow into what we write? I have to ask this as I re-read *Farming Soul*. My path has been defined by these experiences, and as I proceed, veils lift, and I see more. It is not the path I thought I was choosing when I entered analytic training. In fact, it is a path I avoided

---

[1] "Liminality," Wikipedia. http://en.wikipedia.org/wiki/Liminality.

[2] Henderson 2005.

for fear of being deemed ungrounded—or worse. Yet in receiving what was presented to me, it is the path that was laid down before my feet—a path not unlike that of C. G. Jung when he entered his own *dark night of the soul*, documented in *The Red Book*, and finally published in late 2009, a few months before *Farming Soul* came into print. Perhaps only now is the *Zeitgeist* ripe for the telling of such tales.

Individuation demands the balance of feeling within our hearts that we all—humans, animals, the tree across the meadow or the street, the soil beneath us—are of one piece (spirit), while also developing discrete awareness (soul, and subtle body). I have come to realize this is a consciousness of interconnectedness, knowing our survival on the planet is dependent on respecting the paradox that, at once, we are only one part of this complex whole—and we *are* the whole. It is my growing experience of relating to the Other, whether that be to another human soul or the lavender on our ranch here on the western range of the Napa Valley. I suspect it is also a consciousness into which we modern humans are being initiated with the possibility that we become mature, responsible citizens of our Earth and Universe.

Introduction

Fifteen people are seated around a very large, round table in an otherwise empty living room of an old San Francisco Victorian. Norma T., a robust and dramatic woman in her 60s who sits directly across the table from me, is in charge. She instructs us to put our hands palm down on the tabletop. Although she says that she is in a trance, she looks present and alert to me. She could be teaching math, given her matter-of-fact demeanor.

The skeptic in me scans the room. Who comes to a séance, anyway? Most of the people look ordinary: a petite woman with short white hair; a business man in a suit; a young man with a backpack next to his chair; me, a candidate in training to be a Jungian analyst. This could be any-committee-meeting-USA, I think, or a Sunday school class— that is, *until the table starts jumping.*

At first I am annoyed. I suspect someone is bumping a table leg with a knee. But as the tempo increases and Norma comments that the energy is particularly strong this evening, I realize that the table's legs are actually leaving the floor at times. I am shocked. As if from a long distance, I can hear Norma channeling to the woman with the short white hair. A dead relative is apparently giving her some kind of guidance. The skeptic in me is having a heyday, but my attention is riveted on the phenomenon of the bouncing table. I feel as if I am going to throw up. The room is spinning.

Norma finishes with the woman and calls my name. She will read for me next. She seems to know my condition and says, "You become nauseated when Spirit speaks to you." I try to center myself. I do not remember what she tells me, except that I need to wear my hair in a less ordinary style, a comment that insults me. What I remember most is my altered sense of reality.

The bouncing table challenges my ideas about how the physical world operates. It has taken me years to work through my resistance and get to this strange gathering, and it will be two more before I am ready for any real instruction from Norma T.—instruction, as it turns out, that is necessary for me to heal old wounds and move on in my life.

* * *

In 1989, I began training to become certified as a Jungian analyst. I knew the process would be rigorous, but I had no idea that by the time I finished 11 years later, almost every aspect of my life would be irrevocably changed. The forces that reshaped me included my involvement in a shamanic study group, an intensive nine-month study with Norma T., and my education in the techniques of biodynamic farming—not to mention my conventional coursework and personal analysis. All became essential to both my survival and the completion of my certification. This book tells the story of that time.

Some analysts and candidates have found that the Jungian candidacy is an initiatory process. It certainly was for me, resulting as it did in approaching my life in a new way. Such a process operates under the archetype of shamanic initiation: a period of suffering, including death and dismemberment of the ego's old relationship to the Self [3], and eventually developing new "organs." One then returns to her life as an initiate, having been born into another kind of consciousness. Every time this cycle repeats, the process—also familiar to the artist and writer—is more firmly established within. Eventually, in the words of the alchemists, the stone of one's self begins to crystallize. My initiation process was marked by pain and suffering, but also ecstasy.

* * *

The narrative begins in the present and moves in circles, braided into three strands: one, the annual cycle of biodynamic farming on our

---

[3] Jung used this term to denote the central organizing principle of the psyche, mostly unconscious.

ranch in the Napa Valley; the second, my path to become a certified Jungian analyst; and third, my transformational work with a variety of teachers and healers.

I was not instantly transformed in a brilliant flash of light. I have found, both for myself and in my work with the patients whose journeys I have been privileged to share, that most initiations are not sudden and heroic at all, but happen only after long and careful reflection, so this telling is also part of my initiation, as I comprehend and integrate my experiences more deeply, more completely.

Sometimes the writing of my life story resembles watercolor painting. I go over and over it, each wash lending a slightly different hue. In each telling, old patterns recede as new colors and forms emerge.

Storytelling weaves both teller and listener into a larger fabric, suggesting correlations and deepening understanding which opens us up to aspects of ourselves that we override in everyday life. Not that our logical minds have too much to do with it. Something beyond words and thought is summoned, and we must surrender to the process.

At the beginning of each session with Norma T., she offered this prayer: "Drink of the cup of forgetfulness, to remember who you are." Then, together, we would sip a bittersweet, aromatic tea.

In writing this story, I have drunk of this cup ever more deeply. More has been forgotten, and more remembered, too. As you read may you, too, forget who you are and "re-member" yourself as a part of something greater.

One

Late January

The grapevines are quiet this morning. Earlier in the week Natalio reported that the vines were waking, sap leaking from the fresh pruning cuts. But last night it froze hard. Thick, white frost still outlines the grass growing in the shade of the mountain. This morning Natalio says the pruning cuts are dry. Even in the sun, he wears gloves for warmth as he works. Although I can hear Natalio singing in his native Spanish, the atmosphere is quiet—cold and quiet.

The ground is muddy from the solstice rains and the intermittent showers that have come ever since. A jackrabbit darts across the path ahead of me, this way, then that. I pull my jacket tighter and let myself settle into the energy of the vineyard. I, too, feel quiet, receptive, relieved that it is not yet the time for action. Our biodynamic consultant, whom we call B.B., described this time as the crystallization period, those fallow months when the earth is most receptive to the cosmic forces. I savor this knowledge as I walk. I imagine Orion's song, the chorus of Pleiades, harmonizing with Natalio's voice, and the roots of the vines listening intently.

Seven months earlier, we had buried my mother's ashes, at her request, mixing them with my father's. We gave what was left of my parents to the rich, waterlogged humus of the Illinois prairie that three generations of my family have farmed and buried their dead in. That day of the summer solstice was hot, and as we poured the ashes of my mother into the grave, mixing them with the 10-month-old ashes of my father, a breeze swept the particles into a spiral, rising out of the grave, raining them onto my sisters and brother and me. I felt the grit of my parents' ashes—on my face, in my eyes, even in my mouth— throughout the memorial service that followed. The taste was surprisingly neutral. I was once formed out of the stuff of their bodies, and now I breathed, swallowed, and absorbed what remained of them back into me.

That was the summer solstice. Now, walking in the winter vineyard back home in California, the plants' energies withdrawn deep into the soil's embrace during this time when the earth is most open to Spirit, I wonder: Do those minerals I inhaled, ingested, absorbed listen with me to the songs of the Pleiades and Orion? Are my parents' bodies, now particles within my own materiality, still attuned to the cosmos? Is this how wisdom is transmitted?

I descend into the inner world so quickly now! It is the result of living these many months on the edge of death. First my father's sudden passing, then my mother's slower transition over the course of 10 months.

They both lived far away in the Midwest, where I was born and raised on a small farm near a town one-hundredth the size of Napa, the city closest to where I now reside. I still live about three miles out of town as I did then. If you add all the days together, I have lived within city limits no more than a year of my life.

My husband Donald and I live on a ranch at the foot of the Mayacamas Mountains overlooking the Napa Valley. We live on a long and narrow tag of land that used to be a trade route between this valley and the next. The old timers tell us the meadow south of the kitchen was used as a picnic ground for early settlers. Before that, it was most likely a ritual ground for the Wappo Indians.

From our kitchen one sees a grouping of valley oaks. You do not realize the trees form a complete circle until you step inside and the trees join around you. The feeling within the circle is one of absolute presence. You forget time. An hour could pass or a minute—it is all the same. Some people say this whole meadow is a spiritual vortex. Indeed, I often feel that I am living on the shores of a lake of energy. At night I open our south bedroom window, and I can hear the energy waves lapping against the ground near the garden.

I have two young adult sons, Jesse and Casey, whom I love more than my own life. Given a hard choice, I would die for either. Raising them healed me in ways I could never have anticipated. Their father left me many years ago, four months after the beginning of my candidacy to become a Jungian analyst. It was one of the greatest griefs of my life at the time, but I recovered and remarried. These days some of my happiest moments are with Donald, a retired architect, in the home he designed and we built together, where much of this story takes place.

Two

The Dream Pond

Until I met Donald, I had no intention of returning to farming. I left the family farm on which I grew up when I graduated from high school, as soon as I was able. While other students cried when their parents left them at the college dorm at the beginning of freshman year, I wandered the campus in ecstasy.

I had hated fieldwork and the constant labor of the farm and found small town life suffocating. I was on to better things, whatever those 'things' might be. After a few false turns, those 'things' came to include California and psychology, while my relationship to the earth became more that of hunter-gatherer—not tethered to a particular plot of ground every waking moment of every day.

Eventually I married a man tied to the land, an architect with a vineyard in California's Napa Valley, and my sons and I came to live on his property. One braid of this story is about my redeveloping a relationship with the land, one rooted again in a particular place and guided by the tenets of Rudolf Steiner's Biodynamic agriculture. A second braid is my professional path to become a Jungian analyst. Surprisingly, these two seemingly unrelated braids in my life are deeply intertwined.

A most useful tool of a Jungian analyst is the dream. Dreams reflect the spiritual life of the psyche and often compensate for what we consciously suppress, or reflect what has not yet come into consciousness. Those dreams that announce the beginning of a new cycle in an analysis or a life are particularly important.

My first remembered dream repeated itself several times in my early childhood. I must have been about three the first time I had it. *I wander out over the grassy field west of our country church, the area my parents would later choose for their gravesite, and the place my brother, sisters, and I would bury what remained of their ashes. I walk as far as you can see from the church, and then I enter into an area dense with vegetation where I have never been. A rich, fungal scent permeates the air. Nestled in a slight dip in the land is a small pond. Two mallards float in the water, which is as still as a scrying mirror.*

The most remarkable part of this dream is the feeling that I am inseparable from all that surrounds me: the ducks, the pond, the dense vegetation, and the horizon. My body is still my three-year-old body— yet it stretches to the horizons. There is a feeling of great peace and wholeness.

As a child I would awaken from the dream compelled to search for the pond in the waking world—not because I thought it could be found, but because, while searching, the feeling state of the dream intensified, and I felt at one with my surroundings. I also experienced this feeling when staring west into the cornfield at twilight, the stalks tall and tasseled, pollen thick in the air—a sense of portent that would take years to understand.

All my life I have treasured my awareness of two realities: the outer experience, in which we all live most of the time, and its rich inner counterpart. In the best of circumstances, like notes in a chord, these two worlds harmonize in a way that enhances the experience of both realities.

Some psychological theories describe a move into the inner world as defensive, an adaptation some human beings develop in order to cope with unbearable aspects of life. This could be true, but my own experience has taught me that the urge to develop an inner, spiritual life is a healthy and often imperative impulse. Certainly this was instrumental in drawing me to the philosophy and teaching of C.G. Jung.

Jung said, "...*the psychological problem of today is a spiritual problem, a religious problem.* Man today hungers and thirsts for a safe relationship to the psychic forces within himself. His consciousness, recoiling from the difficulties of the modern world, lacks a relationship to safe spiritual conditions. This makes him neurotic, ill, frightened."[4] Jung's life work revolved around this point, impelling man to understand that God is working to become conscious through man, and that the meaning of the Christ had to do with the indwelling of spirit in matter.

Western psychology, and Western thought in general, tends to ignore or actively reject the notion of spirit. Sadly, in recent years even Jung's analytical psychology often does not *embrace* the "spiritual." When I use this term, I am talking about the inner realm perceived through meditation, prayer, active imagination, or simply turning one's attention inward, as people have done throughout the ages and in every known culture. There we may experience images, sounds, and bodily sensations, feelings that have special power and meaning for us.

Some of my analyst colleagues have become phobic of the spiritual perspective, fearful that they will be accused of being "ungrounded" or even psychotic. Even Jung himself has been labeled psychotic by some in the psychological community. This has prevented many analysts from intimately knowing and acknowledging the reality of the inner landscape that is, by its very nature, spiritual. In fact, one of analytical psychology's inheritances is the essential work of regaining psychological balance through reconnecting with one's inner spiritual resources.

For me, receiving training in accessing and understanding non-ordinary states brought a much larger perspective. My psychic teacher, Norma T., often said to me, "You are only partly here; most of you is out there..." Jung talked about this in other ways. He described man as having two souls—the impersonal, or ancestral, soul he was born into this life with, and the personal soul that he develops in this life. The newborn's mind is, Jung says, "a finished structure...the result of innumerable lives before his and is far from being devoid of content."[5] He described these two souls in man as being often in direct opposition, dreams and active imagination often reflecting the conflicts between them.

---

[4] Jung 1977. p. 68.

[5] Jung 1977, p. 57.

Jung asserted, "All dreams reveal spiritual experiences, provided one does not apply one's own point of view to the interpretation of them."[6] Interpretation involves thinking and differentiation, prominent aspects of the masculine principle. Listening to spirit requires a very different approach, one into which I was unwittingly initiated during my years of candidacy and primarily through experiences outside the actual training.

My childhood dream seeded me with an experience that our culture as a rule does not support. The dream was so compelling that I could never forget it. Growing up on a farm, spending long hours alone or playing with my sister outdoors in the natural world, sprouted that seed. I began to experience non-ordinary reality early on, but until I developed tools to navigate this other realm and my ego grew strong enough to use those tools, these experiences remained split off from the rest of my life.

* * *

The disciplines and teachings of two men have helped me heal this split in myself, a lifelong project! When I first read Jung's autobiography, *Memories, Dreams, Reflections,*[7] I felt I had finally found someone who experienced the world in some of the ways that I did. I have written more fully about this in a book Naomi Ruth Lowinsky and I co-edited, *Marked by Fire: Stories of the Jungian Way.*[8] Much later when we had a farming crisis and my husband and I were initiated into biodynamic farming, I was introduced to the philosophy and teaching of Rudolf Steiner.

Johann Wolfgang von Goethe was the great granddaddy of both Jung's analytical psychology and Rudolf Steiner's Biodynamic Agriculture. A German poet and philosopher living at the turn of the 19th

---

[6] Jung 1977, p. 71.

[7] Jung 1965.

[8] Damery, Lowinsky 2012.

century, Goethe is best known to the psychological community for his book-length poem *Faust*, but he also made significant scientific contributions, especially in botany and the science of color. Much of his insight came via intensive direct observation and a resulting consciousness similar to my dream pond consciousness. To know something, he proposed, means to hold all aspects of it at once. Only then does a complete picture come into focus, a kind of unity consciousness. "If you would seek comfort in the whole," Goethe declared, "you must learn to discover the Whole in the smallest part."[9]

One of the first editors of Goethe's scientific work was Rudolf Steiner, born in 1861, almost 30 years after Goethe's death. In 1889 Steiner was hired to edit Goethe's scientific papers. As he pored over the great man's work, he became convinced that Goethe's way of knowing produced a state of consciousness quite distinct from the mechanistic worldview widely accepted even to this day. For the rest of his life, Steiner worked to develop and teach this "new" consciousness, which he recognized as the direct perception of the spirit world that he himself experienced from an early age.

Biodynamics is the agricultural extension of this way of knowing. Using his natural clairvoyance, Steiner studied the etheric formative forces influencing the life of the land and plants and devised ways to balance them. In keeping with Goethe's thought, he observed that there is nothing in nature that does not bear relationship to the whole. The work for the farmer, then, is not killing the aphids that have suddenly arrived on the lettuce, but studying the conditions that resulted in the appearance of aphids and using developed thinking to comprehend the connections between them. This process, done correctly, produces a state of consciousness similar to what I experienced in my pond dream.

Carl Gustav Jung II was born in 1875, the son of a pastor. It was rumored that Jung's namesake and grandfather, Carl Gustav Jung I, was conceived in a liaison between his great grandmother, Sophie Jung Zeigler, and Goethe himself. I suspect that Jung believed this and felt a

---

[9] Naydler, p. 59.

deep kinship with the philosopher poet.

Jung found special significance in Goethe's *Faust* poem, which depicts the process of redemption and individuation through a painful holding of opposites. In Goethe's version of this myth, Faust grapples with his inner darkness through the figure of Mephistopheles and, in so doing, develops a new conscious attitude. Jung felt this attitude offered redemption through a recovery of the feminine principle.

I still vividly remember sitting in my country church as a child, the P.A. system buzzing as the pastor drones on through his sermon. Through the open west windows comes the laboring hum of a tractor, perhaps a quarter of a mile away. Birdsong is sweet and persistent. Even with the minister's amplified voice, I can distinguish the melodic yet harsh song of red-winged black birds, the coos of mourning doves, the Bob…Whites!

Feet shuffle. To my right, my father's napping form slumps against the hard oak pew's splintery smoothness. The June heat is oppressive. I sit very still. The grassy smell of freshly cut hay wafts in the windows.

Suddenly, I am aware of a visitation of Presence! It's as if the Holy Ghost has descended into me and is viewing the sanctuary through my eyes, hearing the droning voice of the minister through my ears, feeling the sweat break through the pores of my body. I am also aware that this Presence is a larger part of me that is seldom noticed, and the moment it becomes conscious it fades away.

I see this as an experience of the feminine principle at work, more heart-centered, emotional and intuitive. In contrast, had I experienced this moment in church primarily via the masculine principle, I would have used only my rational mind, listening to the words of the minister, comprehending the principles he was advocating, and applying them logically to what I should be thinking or doing in my life. This activity, quite useful when it is kept in balance, is centered in the head rather than the heart.

In the past 300 years, most human societies have considered the masculine principle to be more important and more valuable. We perceive the feminine principle as lacking import and being nonproductive; we relegate it to the purview of the mystic.

In truth, we need both ways of experiencing reality in order to live whole and balanced lives. Embracing one and precluding the other leads inevitably to conflict, within ourselves and between others.

* * *

In his autobiography, *Memories, Dreams, Reflections*, Jung described the Faust myth as his own. He identified with Faust's fate, saying it "awakened in me to the problem of opposites, of good and evil, of mind and matter, of light and darkness."[10] Through Faust, Goethe offered Jung "a basic outline and pattern"[11] for holding his own internal contradictions and developing his philosophies, which we study to this day.

It was Jung's work that first provided a matrix for understanding my own experiences of Presence, or so-called non-ordinary reality. This was the only reason that, in my early twenties as a graduate student of psychology, I was drawn to Jung, and to his work on alchemy. In no conscious way did I understand at that time what I was reading, and yet, on some deep level, I felt completely satisfied.

As I matured I wanted to understand the experiences and the images that appeared in dreams and visions. It was then that, unconsciously and then consciously, I sought teachers.

---

[10] Jung 1965, p. 235.

[11] Jung 1965, p. 235.

Three

My Early Professional Life and Personal Analysis

For ten years before my application for candidacy at the Jung Institute, my life revolved around two and then three hours a week of Jungian analysis with an analyst I will call Benjamin. If you have never experienced an analysis, you may not know the level of commitment and involvement this implies. In addition to the hours spent in the room with the analyst, there is travel time, which for me meant a three to four hour round-trip drive. Then there are the hours between sessions that one spends journaling about what happened in the session, one's reactions to what happened, and dreams and insights, as every relationship in your life becomes more transparent. In short, for the first few years of my analysis, my entire life revolved around my work with Benjamin.

In fragile shape I first arrived in Benjamin's office on June 30, 1978, after a series of catastrophes in earlier therapies and training experiences that today would be considered ethical and professional violations. But we of the seventies were unapologetically breaking through barriers without a thought about transference (the unconscious patterns from earlier relationships that are projected upon the analyst), without consideration for the dangers of dual relationships (therapist/client relationships outside of the therapy), and without respect for the boundaries in therapy that are essential for inner work to be successful.

My first therapist, Priscilla, was just a year ahead of me in graduate school. We were both part of what, in retrospect, was a therapist cult in our area, led by a charismatic and disturbed supervisor, Jen. Nothing in my Midwestern background prepared me for the early seventies in California, and particularly for Sonoma State University and a powerful cult personality like Jen. Jen had been hired independently by her supervisees as an adjunct to our training, not by Sonoma State

University. Her *modus operandi* was, in her own words, to lie beside the road in the weeds and "rape" the non-suspecting client with her interpretations as he or she walked by.

I was in my early twenties when I first met Jen and was completely taken in by her power. I worked with her for several years as I finished my degree and became licensed to practice psychotherapy. Many of us felt that hers was the only true way. All other approaches paled in potency. We knew this because she told us so, *repeatedly*, and it took me a full year with Benjamin to let go of this notion. After a diet of heavy confrontation, particularly around any dissenting voice, and the threat and use of public humiliation by disclosure of personal material, many of us were wounded, to say the least, and I was one of the first to know it. So when my friend Harold gave me Benjamin's name—saying, "Get out of this system, it is killing you, and here is a Jungian analyst"—I was ready.

I wore a mid-calf black dress with tiny white flowers to my first appointment, a long pink chiffon scarf wrapped about my throat, and high, burgundy boots. Benjamin was an attractive man in his mid 40s with a tall muscular frame and thick, curly brown hair. His voice was soft and calming. As I told him about the mess I was in, he began a process that would last for several years, by containing my anxiety with sane statements and alleviating my feelings of self-hatred and blame. I recited the litany of faults that Jen, and then Priscilla, had repeatedly accused me of: I was secretive, manipulative, passive aggressive. To my surprise, Benjamin simply asked, "Is that a threat or a promise?" And for the first of what would be many times over the next eighteen years, we both laughed.

At the end of the hour, Benjamin, reflecting on our session, stated that there had to be some energy between us for the therapy to work and that he could see such energy was there. I was relieved but did not reply. I was horrified at his next comment, that he had never seen a person so white, so drained of blood. I had lost a fair amount of weight in the preceding months, forcing myself not to drop below 105 pounds. I also had forced myself to sit in the sun to tan, telling myself that at the ripe old age of 29, sunning was not a good idea, and that this was the last year I would do so. Apparently, the white ghost of myself was showing through anyway.

I worried that Benjamin did not find me attractive enough. It was a tension that would continue for years: the attraction between us feeling dangerous, so dangerous that I could not talk about it. He tried to get me to express my feelings about him or something he said, by asking questions that required no more than a "yes" or a "no." In time, in a convoluted way, this behavior confirmed that I mattered to him, but after my experiences with Jen and Priscilla, I had no capacity to be more direct about my feelings for Benjamin. Much healing would have to happen before I was whole enough to do so.

Nevertheless, after that first session I felt a "righting" inside. I wrote in my journal, "I have this feeling that I do not have to swim upstream to be a good person. He is gentle, kind, smart, and supportive. His words are vague. He can be as spacey as I am. But he *is* intuitive."

Over the next months, Benjamin's listening and training formed a protective circle around me, and once a level of safety was established, I descended into the greatest and most frightening depression of my life. It was as if the full impact of a massive and malevolent storm waited to hit until I was in his care. At night I would go to sleep picturing him as a gold chain encircling me. I felt psychically dizzy and feared for my sanity. Outer events pummeled me.

Having left Jen behind, my referrals from her group shriveled, threatening my private therapy practice. I suspected that I was being used as a teaching example in her consultations and seminars and so worried for my reputation. I was sharing an office with my old therapist, Priscilla. When that relationship deteriorated, I had to buy her out.

During each session with Benjamin I related a new drama: a large hole in the wool carpet of the office after Priscilla left, furniture being taken without agreements. Benjamin was the only voice of sanity in a crazy world. "You can't make agreements with people who do these things; you have to get out."

In January, six months later, I returned early from a vacation to keep a Tuesday appointment with Benjamin. A seasonal storm was dumping torrents as I drove my 78 Honda Civic through water collecting on Highway 101. The skylight was leaking. I recognized that I was in an emergency situation, but at least I was finally getting help.

Over the next months I was able to stabilize my life, separating completely from the community that had spawned me professionally, finding a new office in which to practice, getting another therapy consultant, and totally cutting off relationships with members of the former group. I agreed with Benjamin that not everything can be fixed and that sometimes you just have to move on.

One day on the way home from therapy, about a year into my work with Benjamin, I looked in the rearview mirror to find myself smiling spontaneously, something that hadn't happened in a long time. Then I dreamed about "whitening," the *albedo*, an alchemical stage Jung interpreted as coming after the burning away of ego attitudes or ambitions not related to the larger Self. Finally, thirteen months after I had started with Benjamin, I married Lynn, the man I had lived with for the past six years. My depression had ended.

But the real inner work had only just begun. At last I had the reserves to move deeper into my relationship with Benjamin and to understand the wounds that had opened or been inflicted by my experiences with Jen's group. And the focus moved from the outer crises onto the relationship between us. One day late in that first year, after the usual greeting at the start of our session, Benjamin picked up a notebook and jotted something down. I stopped talking.

"What just happened?" he asked.

Feeling closed down, I said, "Nothing!"

He reiterated, "You started talking. I picked up the notebook and made a note, and you stopped talking."

"It didn't bother me," I insisted.

"Come on, tell the truth!" he pushed.

"No really, it didn't bother me," I said. "I don't know what happened."

"Would you tell me if it had bothered you?" he questioned.

"Not necessarily," I said.

"I can relate to that!" he said.

Again, we laughed. After that session I wrote in my journal: "I don't want to talk about feelings because I don't *know* anything when I do. I feel helpless. I don't want to be vulnerable, like I was with Priscilla. Priscilla was disgusted by my feelings."

Immersion into Jen's group added years of work to my analysis. The rebuilding of trust in others, and more importantly in myself, took a very long time. I came to think of Jen and Priscilla, and the larger cult group, as my dark teachers with even darker gifts, ones who led me to difficult places in myself that otherwise would have been avoided.

Reflecting back to those early years of analysis, I remember the intensity of total absorption in the process. Analysis is not always comfortable and takes a great deal of life energy. For long stretches of time, one session can feel remarkably like the one before. A large part of my own healing was due to the quiet weaving of attention that I got from Benjamin, the beginning of the Tewa "garment of brightness." At this stage, the warp was the relief I felt when deeply understood and held in his attention; the weft was my disappointment when I didn't feel that way.

Through this process, over time, something was woven back into myself. In one session I felt relief in simply complaining, probably about a friend's behavior or the difficulties in my work. When I asked for a response from Benjamin, he was silent for a few moments, then said that he did not know how to respond. That disclosure in itself was all that I needed. I felt whole. About eight or nine years into my work with Benjamin, I began to pray hard for something "higher." Until then, I was always anticipating satisfaction from our sessions and then was either satisfied or disappointed. There was nothing else – no great insight, no profound transformation. For years it had been enough.

Meanwhile, in my outer life, my husband and I conceived and birthed two sons, my psychotherapy practice grew, and we remodeled our home and then proceeded to outgrow it. I made new friends and joined a women's writing group, which became a lifelong refuge for me. And healing *was* happening. I knew this because of the shift I felt in libido that had previously been tied up in the analysis. Benjamin's vacations were no longer devastating to me. Dreams had become important to me, and not just as the eggs that I gathered each morning to take to my analysis. It was during this time that I applied for candidacy at the Jung Institute and was accepted into the training program

Four

A First Teacher

Besides Benjamin, one of my first teachers and mentors in navigating the realms of my dreams and visions was Don Sandner. Don was a senior analyst at the Jung Institute in San Francisco, a psychiatrist who had also spent 16 years studying with a Navaho medicine man. His book comparing Western and Navaho medicine, *Navaho Symbols of Healing*, came out of this study.

In our Institute community Don wore the mantle of the shamanic archetype, which Swiss analyst Guggenbuhl-Craig described as the archetype most informing our work as Jungians. I was strongly drawn to this shamanic work and entered into consultation with Don a year before I applied for candidacy.

It was not easy to get in to see Don. If he did not know you, he did not return phone calls. When I was looking for a consultant, Benjamin recommended that I see Don, something he may have later regretted. For years my personal analysis had been flooded with dreams of snakes for years, and Don was known to be a snake person.

The serpent plays a role in many creation myths, including the story of Adam and Eve. It is only through the serpent's offering Eve the forbidden fruit from the Tree of the Knowledge of good and evil that Adam and Eve came to be expelled from the Garden of Eden. Such disobedience, and then grief at separation from unconscious wholeness, are the beginning of consciousness, all started by (and blamed on) the serpent. In Hindu spirituality the snake is symbolic of our *kundalini*, or life energy, awakening from its coiled position at the base of the spine and moving up through the chakras, raising our consciousness as it goes. It also represents an instinctual way of knowing and the unconscious in both its positive and negative forms.

Because of my many snake dreams and experiences, Benjamin suggested that I write a letter to Don, telling him why I wanted to work with him. Don wrote back, explaining that he did not have time, but would call when he did. Again at Benjamin's suggestion, every month I sent Don a postcard bearing an image of a snake, reminding him that I wanted to work with him. After three postcards, he called. Could I come on Thursday, September 9? Without hesitation I agreed.

Don had two offices. One was on stilts over the bay in Corte Madera, directly across the bay from San Quentin; the other was a dark and hot office in San Francisco. At first I met with him in either office, depending on where he had cancellations. Both were filled with wooden and ceramic snakes, drums, feathers, books. The San Francisco office was like a sweat lodge, its darkness inviting one to develop a kind of psychological night vision; the Corte Madera office was more of an ark, the water dissolving the boundaries between worlds.

Don sat in a rocker directly in front of me with a stenographer's pad, ready to record the dreams of my patients or, sometimes, my own. He was a tall, balding man with a large belly and the shaman's wandering eye. I never knew which world he was peering into. Despite his size, he seemed to float when he walked, as if his feet did not quite touch the ground.

About a year after I started working with Don, I applied for candidacy at the Institute. Upon admission, I dreamed I was in a monastery. *The monks are invisible most of the time and seem weak in some way. Don Sandner and I are helping them. A shower of water comes down and drenches Don. Then, in this other dimension, as a monk showers, all the invisible people come into view, showing that they are present. They say they don't need to be visible to be helpful. They are working to restore harmony and balance.*

When I told Don this dream, he said I would soon find out that the Institute was not a holy college. That was one level of interpretation. But the dream did reflect a problem in my psyche at the time of my entry into the Institute: the monks working to restore balance were too weak. Perhaps the restoration had to do with strengthening spirit life, both in myself and in the Institute. In the dream Don is baptized by the mysterious shower of water, after which I can see into this previously invisible dimension.

I told Don, "Maybe the shower rained on you first because you are the most receptive to the two worlds." For several years he carried for me the spiritual animus, or contrasexual aspect of the soul.[12] As I proceeded in my work with him, my unconscious quite active, I learned to see into the world of the spirit as well.

As Don's interpretation of my dream suggested, my euphoria at being admitted to a training program that turned down many applicants was short-lived. Within two months, the larger Institute was embroiled in several ethics violations. My first-year candidacy group quickly developed such problems that two of our four members would not speak to or look at each other. Our attempts to get help only made matters worse for all of us; the Institute itself was in such crisis and we candidates were scapegoated in a variety of ways.

A couple of months later, in January of 1990, my personal life completely fell apart when my husband Lynn suddenly left me. Analysis can be hard on a marriage and I suspect mine contributed to our parting of the ways. Lynn never understood the intensity of the transference. In retrospect, I am sure he must have taken it personally. I regret the hurt he must have felt when I had so much tied up in the analytic work; even more, I regret the impact of the divorce on my sons. But I had little choice but to surrender to this path of work with the psyche and the unconscious. For his own reasons, Lynn could not endure this.

Never in my life had I experienced such dissolution and chaos in my outer environment. Needless to say, any illusions I had that I was part of a "holy college" were, indeed, dashed.

---

[12] In classical analytical psychology, in a woman, the animus, or contrasexual aspect of the soul, carries the so-called male characteristics often projected onto a partner.

Five

Early March

In the Napa Valley the massive wind machines dotting the vineyards begin whirring in late February or early March, on those nights and early mornings when the temperatures dip into the mid-30s. By now the buds on the canes have softened and begun to unfold into flower-like leaf clusters. Bunches of miniature berries appear, which, after bloom, will become grapes. It is all there— including the buds that will develop the canes for next year's crop.

The lavender mounds are also pushing out tender, green growth. Frost damages all of this. The time of not worrying about cold, water, or mildew is past. For the next six months, until after harvest, there will be a certain tension.

I am disturbed when I wake in the dark to the whining of the wind machines. The whine, which soon becomes the roar of many propellers of airplanes, usually begins around 4:00 a.m., though occasionally as early as 2:00 a.m. Whatever time I hear it, I get up and close the window.

Generally, I like to hear what is going on outside: the telltale ringing of bells from the goat barn not far from our home—soft, rhythmic tinkling if they are eating, more intense rings if scratching, or frantic clanging if there is trouble. If our llama whinnies his alarm or a mountain lion screams, I don't want to miss it. But I also listen for the barn owls, the western screech owl's warbling calls, and the hoot of the great horned owl. Once, one of my son's birder friends slept in our courtyard, identifying the songs of 21 different birds as he woke: western blue bird, bushtit, Bewick's wren, hermit thrush, American robin, yellow-rumped warbler, dark-eyed junco.

I return to bed after closing the window, but I can still hear the drone of the machines. So I picture the grapes and the lavender blanketed with warmth and I pray, hoping the power of prayer works. I do not sleep.

When the eastern ridge across the valley is backlit by a soft, muddied yellow that will soon have a pink tinge, I rise and make a cup of tea, checking the temperature outside. Thirty-nine degrees. Good! Yet in the lower areas of the ranch, it could be several degrees colder.

The sun rises over the eastern ridge as Natalio appears at the kitchen door. "Do we spray today?" he asks. "There's frost by the brush pile." The brush pile is halfway between the house and the vineyard. So I help him find the valerian tincture that we keep packed in peat in the basement along with the other biodynamic preps. I ride with him in the old blue ranch truck to our stirring site near his house near the road. There we have two 60-gallon oak barrels filled with "sun-soaked" water. We empty one barrel halfway and squeeze into it several droppers of the valerian tincture. I face the east, the direction of the etheric forces, and picture the grapes and the lavender surrounded by warmth, for warmth is the force valerian brings. The cold morning air pinches my cheeks as I stir in the herbal elixir. Picturing a blanket of love covering the vines, I ask the life forces to protect the tender shoots and the buds.

The valerian tincture requires 10 minutes of stirring, unlike the full hour for some of the other preps. First I stir clockwise, the water resisting the movement of the stick. I push through its density of water until a swiftly moving spiral funnels to the bottom of the wooden barrel. As the spiral slows, I reverse direction, stirring counterclockwise which throws the water into chaos, the state that Rudolf Steiner says is most receptive to the Divine. As I stir, over and over, I throw the water into chaos that it may receive the essence, the divinity, of the valerian.

Six

Natalio

Sometimes you find teachers where you least expect them. Natalio, the vineyard manager, is one such teacher. He shows us how to farm in the old ways. It is taking us some time to learn, some time and a lot of mishaps. I would never wish those mishaps on anyone, yet without them we would never have found our way here.

In Chiapas, Natalio might be considered a *brujo*, a kind of witch. A good *brujo*, I like to think, but I really don't know if he is all good. I do know that after Natalio started singing to our vines, the leaves grew thick and green and the canopies had to be thinned. They were so lush, the grape crop doubled.

Of course, it may have been the biodynamic preps, too, and the practices Donald and I learned from the biodynamic consultant Natalio calls *Brujo* Bill (or B.B.), but I have seen the vines tilt toward Natalio as he walks the rows. He calls them his ladies and loves them as children. Once, when an implement salesman accidentally cut an adolescent vine to the ground while demonstrating a hoe plow, Natalio cradled the vine in his arms and cried.

Natalio spends almost all day everyday with the vines and the lavender. In fact, he spends more time with them than with anyone else, even his wife and six children. He knows every square inch of the vineyard.

Several times a day he comes to me. "Are you busy," he asks. This question annoys me. "Of course I'm busy," I respond.

"You are always busy," he answers. "There is never a time that you aren't doing something."

I know he is right, but this annoys me even more. "What do you need?" I ask rudely.

"Can I show you something?" he counters. Impatiently, I get my hat and follow him out the door.

He walks several paces ahead of me to the vineyard. The sun is warming up these late winter days. He walks the vineyard edge of wooded wild lands where the deer sleep. One of the coastal oaks along the fence has a gaping hole in it that is alive with bees.

Turning to me, he says, "I am watching this." I know what he means. In part, it is a rebuke for the swarms that Donald and I passed up over the years. Then last year, when we finally decided to keep bees and bought two swarms, one immediately dwindled to nothing. Natalio was furious. The right kind of bees for our land were the swarms that came to live here of their own accord. He had seen one in the tree two years ago. It waited a week for us and we did nothing. Then he saw another by the creek last spring. Again we did nothing. When we finally thought we were ready and a woman in town offered to sell us a couple of swarms, we bought bees that Natalio considered weak.

"These are healthy bees," Natalio chides, showing me the natural colony. "They return with much pollen on their feet." He shows me the highway of bees flying east and returning, fast-moving golden darts illuminated by the sun. I step closer to get a better look at the hive and see pieces of jagged honeycomb jutting up near the mouth of the hole. A bee bumps into the side of my head. "Watch out!" Natalio says. "Don't get in their way! These are good bees. We don't want to train them to sting us."

He talks to the bees, calming them. When they bump him, he stands frozen, waiting. The bees return to their collecting; we turn back to the gate. The sun is sinking below the tree tops. 3 p.m. Soon the bees will be heading home, Natalio says.

Once we noticed that the bees in our purchased hives were agitated. Natalio's eyes shifted about. "They are angry," he said. "There is a lizard."

And sure enough, a lizard was climbing the concrete block footing to their hive box. Natalio worked for some time to fix the opening so that the lizard could not enter, though he probably would have been no match for the bees.

Ronda, the woman who sold us the bees, learned beekeeping at the local junior college. She suggested that Natalio take a beekeeping class to become current on bee diseases. Natalio does not read or write, nor does he want to. His eyes were unblinking as he listened to Ronda. Suddenly, he no longer understood English! But later, in disgust, he told me how little respect she had for the bees, entering their hives too often, bringing bees in from other properties, using smoke to subdue them.

One of Natalio's faults is that he talks too much. Everyone thinks so. Donald says that when the two of them shovel manure out of the flat bed, Natalio spends half the time leaning on the shovel telling stories. Natalio tells how he narrowly escaped marriage to his boss's daughter by leaving town when he was a cook in an Italian restaurant in Kenwood, how he almost died as an adolescent having a surgery that none of us can understand, even though we have heard this story many times. How his grandfather lived to be 120 by eating *nopales*, the same cactus that he planted in our courtyard and that Donald and I prune only on weekends when Natalio is gone. Once when he found a pile of its thick leaves in the garbage, he accused us of murdering it.

Natalio tells of rough times in Mexico when he was growing up, of the need to carry a gun and "take lives" if one was to have respect. He says he had 100 hives of bees and as many cattle. He says one cow quit giving milk and his grandfather told him to follow her into the bushes and watch. Sure enough, he says, a python came, climbed the cow's leg, and nursed. He told me that if a baby loses weight, the python is probably coming in the night and stealing the baby's milk. I laughed when he told me this, but he did not.

We have our theories about why Natalio talks so much. Donald thinks that he is lonely and is relieved to have human company. This may well be part of it. It may also be Natalio's way of slowing down, respecting the body, pacing himself for all the years he knows he will be working day-in and day-out. But the telling of stories also has a useful purpose, bringing us into that frame of mind that is very much a part of farming in the old ways.

Watching the bees and waiting for a swarm, knowing by observing their work in February that a queen is being created, not interfering with natural processes, intimately knowing your plants, your field, your land through observation — these activities foster a state of mind in which the eye and ear of the imagination open. It also is the way animals communicate, Norma T. told me once. Being on the land opens you to that.

Seven

Candidacy and Initiation

In the midst of the chaos of my early years as a candidate to become an analyst, and perhaps in reaction to it, I took a trip that would forever change my life. In our second year, a small group of candidates asked Don Sandner to take us to the Corn Dance in Santa Fe the following August. He agreed, and on August 3, 1991, ten of us met in Santa Fe to attend the Corn Dance at Santa Domingo Pueblo the following day, and then to tour the Anasazi sites in the area over the next week.

We arrived on a Saturday afternoon, converging in a condo that we had rented for the weekend. The group was composed of six candidates, two spouses, Don, and a friend of Don's from Colorado, Steve. I had a terrible headache that day. The excitement and altitude had taken their toll, so I went to bed while the others went to dinner. Although I was in pain, I felt excited and full of expectation. The evening was oily black, everything shiny and slick with rain, the asphalt glistening in the street lights.

I could spend a lifetime writing of that experience: about the sparseness of the New Mexican landscape, the smell of pinion pine and the stark beauty of the turquoise sky, buildings made of logs encased in earth, the dry air and afternoon thunderstorms that dumped torrents of rain. I could write about the way time opened during those days, an experience I previously had known only at twilight or when I searched for the dream pond as a child. It felt that I was in a rare and precious gap, and that—if I stayed present to what was happening, if I took it all in—nothing would ever be the same again.

That first night, I dreamed that I and the rest of the group were traveling and our skin turned leather brown from the sun. I also dreamed that Ellen, my roommate on the trip, and I were on a trail through a volcanic region full of fumaroles. In the dream Ellen warned me to stay on the trail or I could be burned. I felt warmth from the mud emanating around me and stayed on the trail to avoid injury. As I was to learn in the coming years, it is indeed wise to take precautions when dealing with great energy.

When I woke the next morning, the world fresh from the rain washing and the sky clear, my headache was gone. The pain of the last couple of years, which had cracked me open like frost cracks the husk of a walnut, was finally letting up. I was ready to receive.

The Corn Dance had all the trimmings that make me steer clear of such events: long lines of traffic to get to the Santa Domingo Pueblo, crowds of people, standing room only. But immersed in a more open state of mind, I was oblivious to what normally would have felt intolerable.

When we arrived at the pueblo that morning, the dance had already begun. In the distance we could hear the steady beat of the drum. Kosharies, white mudded clown-like figures with black spots and corn husk hair, ran about being obnoxious. "Don't get in their way," we were told, "they can be rude!" And rude they were, their behavior considered outside the law for the day. If unknowing or inconsiderate tourists forgot or ignored the rule of no cameras, the kosharies just took the cameras and did whatever they pleased with them, emptying out the film in the process. They teased, insulted, and kept a kind of absolute order, not unlike the Grim Reaper. No bargaining! The kosharies danced to their own beat, not in line with the ritual dancers, but outside and weaving through the others, now kindly adjusting the fox tail worn by a four-year-old, now chastising someone for getting too close to the action.

The dance itself was a great convergence of energy. Never had I been in a place where so many people were dancing in the same rhythm. The Native American women wore black dresses with one bare shoulder and carried evergreen bows. The men were barefoot and in traditional dress, with giant bells strung on their belts and parrot feathers pinned in their hair. The rhythmic ringing hypnotically accentuated the drumbeat.

The people of the pueblo were divided into two halves, or moieties: summer, the Turquoise kiva, representing the masculine principle; and winter, the Squash kiva, representing the feminine principle. First one group would dance, then the other. Summer followed winter which followed summer, all day long. The drumming and clanging of the bells were ever present. At the end of the day, the two moieties, in seeming competition earlier, came together as the culmination of the ceremony. Then the rain came, for this dance was a prayer for rain and fertility and growth of the corn.

And rain it did! On our way back to Santa Fe that evening, we drove through foot-deep water. Suddenly, to our left, a bolt of lightning struck a telephone pole and a bright silver-blue ball of fire hung in the air for a few seconds before dissipating. In my altered state of mind, I wasn't surprised by the spectacle, nor were my companions.

Over the next week, we visited prehistoric sites by day and by night drummed and listened to Don present his material on the Navaho sand paintings. With this kind of immersion in unconscious process, one can take in this imagery in a way that our fast-paced Western way of life precludes. Like the pueblo dancers of the opposites, who in time become complementary parts of the whole, each day we, too, swung between direct wordless experience and rational training.

The evenings began with Don smudging each of us with white sage smoke to cleanse our energy bodies and prepare us for the psychic work of drumming. Next, we each found a private place to relax for an hour. Don read a prayer or poem, then he drummed solo for an hour, his beat twice the rate of the human heart.

The first evening of our week-long stay in an aging but cheap motel in Ojo Caliente, the proprietor allowed us to use the closed dining room for our drumming, and there we found our various spots. As the drumming began, Don read a prayer by Kabir. "Between the conscious and the unconscious," his voice boomed, "the mind has put up a swing." Images formed and his words called me back: "All earth creatures... sway between these two trees, and it never winds down." His voice continued, "Everything is swinging: heaven, earth, water, fire, and the secret one slowly growing a body."[13]

The secret one slowly growing a body. The words penetrated me, coming from a distant yet intimate place that I knew, yet had never heard of. Later, I would read Barbara Hannah's essay, "The Beyond,"[14] about the formation of the subtle body, and Norma T. would name the anatomy of this "secret body" for me. But this night, under a New Mexican sky studded with stars, the words simply ripped me open for what was to come.

The drumming continued. As I listened, between the beats and beyond them, I heard a flute. Allowing my voice to softly imitate the sound, joy came and I was filled with ecstasy. I became the flute and the wind was the wind before a thunderstorm, howling, cold, and blowing right through me. Sometimes I felt filled with fire – wind and fire. The flute opening was my own throat. I was the plain, I was the wind, I was the sound: so cold, round, and appealing. I wondered with awe if this is how it felt to be dead.

Not surprisingly, I had a hard time sleeping after this experience. Finally I got out of bed and cloistered myself in the small bathroom of my shared motel room so as not to disturb Ellen. Sitting on the bathroom floor, I painted the flute on a pad of watercolor paper. I tried to sleep again. When I finally got up for good at 4:30 a.m., my vision had disappeared. The jagged geometric pattern announcing a migraine was all I could see. I told Ellen that a Navaho rug pattern was dancing before my eyes. I sat in meditation and soon my sight returned to normal.

* * *

That day we were to meet Don and Steve in Abiquiu before driving on to Chaco Canyon, a six-hour trip. I would be at the wheel. While driv-

---

[13] Bly, p. 18.

[14] Hannah, pp. 36-53.

ing across the plains, barely lit by dawn, I forgot to look at the gas gauge. When I realized that we were almost out of gas, I parked alongside the road and we waited for Don and Steve to drive by on their way to meet us. Still on my ecstatic high, I read Indian prayers aloud. I felt energy rushing in my body like a mountain stream at thaw. It was the energy of the cosmos, *kundalini*.

The *kundalini* awakening is a humbling experience. First there is the ecstasy. I luxuriate in that soaring, but as the experience proceeds, the increased energy shakes everything loose —every doubt that I ever had, every rage. In short, one's *kundalini* on the rise rattles the complexes — everything not related to the Self, everything not nailed down. This experience is the Tower Card in the Tarot, or, in alchemy, it is called the *nigredo* that comes after the lesser *coniunctio*, that early, energetic coming together of incompletely differentiated aspects of the Self. Things fall apart; ego structures not related to the larger Self crumble.

When this happens, I barely endure it, remember not be overcome by the complexes or act out by being grumpy or insulting to others, or falling into a heap of self pity. I try to use the time to examine myself and face what is coming. Otherwise, I risk growing back into the worst of myself.

That day on a road in New Mexico, I was certain that everyone was angry with me for the car running out of gas (which they may have been!) and then judgmental of me for reading the Indian prayers beside the road while we waited. I felt stupid for buying a rug that looked like a migraine vision pattern and humiliated for what I figured was a positive father transference to Don. This mental chatter went on for most of the day. Then, as I sat in the kivas at Chaco Canyon feeling the cool of the stone and the wholeness of the circle, the energy calmed. Several of us sat for an hour in the large kiva. I felt all of myself then, immense. The storm had subsided.

* * *

Each day we visited a different prehistoric site, listening as Don told us what he knew. His summers studying in the Southwest with Natani Tso meant he knew the area well. At lunches and dinners he talked, and we hungrily took in his wisdom. One of our group likened us to a flock of little ducks following our mother hen. I think that our group truly got the best of Don's teachings and wisdom, both on that trip and in the next few years.

Midway through this Southwest trip, our group climbed the cliff above Ojo Caliente to an ancient pueblo site that another candidate and I had scouted out earlier. We had located the site mostly through educated intuition, having learned the layout of the pueblos from visiting various ruins and knowing that there was a pueblo on the cliffs above our motel. As our group started the climb, we saw a dried snake hanging on the fence. I instinctively screamed and jumped, inadvertently jiggling a ceremonial rattle in my backpack. We laughed all the way up the cliff.

The drum was heavy and big, about the size of a backyard grill, and the men carried it by cradling it in a chenille bedspread. Once on the bluff, we proceeded along a path to the ruins of the pueblo, along old water ditches dug by people who lived hundreds of years ago. Again I felt open to the land, like I had felt in the childhood dream, and a knowing came: *Here is the pueblo, here are three kivas, here is the right place to drum.*

There was much laughing and complaining about the elevation of the climb, the distance, the heat — until we started drumming. Immediately, the drumming brought silent reverence. Again Don read an invocation, a Tewa prayer poem that he had read many times before. As my eyes rested on the distant horizon, deep recognition grew. This was the very landscape which had inspired those words. I heard the words as if for the first time, as if they were being inscribed on my soul. *Then weave for us a garment of brightness that we may walk fittingly where grass is green …* [15]

The noon heat opened the sky, the earth, my father's fields to me. It was so hot, so very hot, that my throat opened, and I *felt* my voice. I had to trust my vision. I was worried that two hours in this heat would make me sick, but I felt something rise up within and knew that the heat was my friend.

There were presences there. During the drumming my hat blew off, or rather someone took it off, using the hand of the wind. *Take your hat off!* I heard. Then there were warm hands upon my head. I looked up and Don looked at me. In another realm, a rattlesnake coiled in our midst. Later Don told me that he saw the snake, too. I felt consecrated—not ordained, which has to do with order and hierarchy, but consecrated, "made sacred," or, as Steiner would say, reconnected with my underlying spiritual power. [16] I remember feeling that I no longer needed a priest (my analyst) for a relationship to the unconscious. Something was changing in me in a profound way, and as a result my body was *consecrated* as a holy temple.

That evening we drummed again, and this time in the vision I was led into a low, dark, rectangular hut to write. I was tended by invisible presences; it was cool and deeply restful. When the drumbeat began, I again heard the flute. Its clear notes were present to me throughout the trip. To this day, whenever I hear flute music, I am transported back to the spare New Mexican landscape.

At dinner Don continued his teaching. He said that the transference to the Self comes after the parental complexes are mostly analyzed. The animus/ anima are the mediators to the Self and the source of healing powers. "Once you've been healed in this way," he said, "you are obligated to be a healer."

I had an archetypal transference to Don, quite different from the demands of the personal transference that I had had to my analyst,

---

[15] Spinden, "Song of the Sky Loom," p. 94.

[16] Baan, p. 72.

Benjamin. The day after the drumming on the bluff, I woke from a dream in which I learned that a loved one had died. In the dream were the words, "*First there were four, then there were three.*" Because of several associations to this person who had died in my dream, I immediately interpreted this as the death of my personal analysis, at least in its regressive period. Perhaps it was the growing conscious connection with Don that facilitated this shift.

Yet I was overcome by grief, knowing something was irrevocably over. I prayed that my grief reverberate through me, like the wind through a flute. *Play me, I am yours*, I prayed. *Be with me in my sorrow that I may live. Be with me in my joy that I may survive it.*

I became aware of the presences and again felt that I was being shown things. Were these presences aspects of my psyche, or did they have an outer, separate existence? This is a big question. Before I had only felt them when I was despairing, and then they were my grandmothers and great grandmothers. On this trip, things changed. They were not just my grandmothers, nor were they only female. I didn't know them, but they came as teachers and helpers. Clearly I needed to learn to live in a way that supported my feeling them, perceiving them.

These presences were particularly available during the drummings. I entered sacred space by making the sound of the flute in my throat. Humming the sound, *being* energy, became synonymous with feeling the presences. They told me to go deeper and I did, sinking into the earth and feeling the strong energy enter my feet and push up through my body. I was a large black snake, or rather a large black snake traveled through me, reared through my body over my head, and I felt it was me—my shoulders gone, my face a snake face, my mouth a snake mouth, my eyes green snake eyes. I thought "This is too much!" and pushed it away.

"It's okay," the presences said. "You'll get yourself back."

Maybe so, maybe not, I thought. I was scared, not sure what to do next.

The presences said, "Don't worry, it's only practice. It's like childbirth, this thing of moving energy through your body. Let go and let it happen, this other force, not you."

Then I breathed deeply and I was flying, flapping my wings up and down.

* * *

The last morning of the retreat, I climbed up the hill, returning to the old pueblo. I was aware of the gifts of the trip, and of Don, who had given us his time, energy, and teaching. The presences were near. I felt them in the air, almost saw their essences. They were men this time, Native Americans who knew agriculture like my father's family knew agriculture. I felt they were there to heal me, to help me learn to heal. They told me that I needed to stay open to them. If I did, they would help me.

Each of us spent that last morning alone before coming together for a final drumming at noon. I walked to the creek that ran along the base of the bluff and found a space to write and paint. The presences were very near. As I sat there along the creek, those healing waters containing lithium and arsenic, the Snake came again.

"Paint your face in the mud," it said, and I did.

Then, frantically, I wrote as the Snake dictated my healing song. I sang it aloud as I wrote the words. Its melody is the landscape of my soul, and to sing it always brings me directly in touch with my psyche's deepest resources. When the Snake finished dictating, it instructed me to dampen my watercolor brush in the creek and paint the Snake itself. The picture is stunning, full of awe. It is perhaps a serpent's head, perhaps the *sipapu*, or hole of emergence, into the next world. As I let the paint dry, I rinsed my face in the creek's waters.

* * *

Our group's final ceremony seems so long ago now. Afterward, everyone but three of us left within fifteen minutes to catch planes or fulfill promises. I went into immediate withdrawal, an animus attack, fighting depression and, again, every complex that my psyche could conjure. I suffered it consciously, naming it for what it was: the after effect of tremendous energies moving through me, offering me the opportunity to be larger than I had ever been, and the resultant sloughing off of all that no longer fit.

Don said the drum is the Self, ever present, pulling you back. Those days were one long exercise in the ego relating to the larger energies of the Self. In the vision during the last drumming, the Snake curled itself around me, extended itself, and looked me in the eye. It said, *Develop a spiritual practice*. I was given direction for the next part of my life and told not to worry too much about the Jung Institute. I was told to stay related to the Self as I became an analyst, to not be eaten by some ambition to become anything other than myself.

Don brought closure by sprinkling bee pollen on each of our tongues. Don, as I said, was a large man, fond of meat and cake, so it wasn't easy for him to kneel before each of us and ceremonially sprinkle the pollen into our open mouths. I can still see him in my mind's eye, wearing a blue plaid shirt that almost doesn't button around his large belly. His hooded eyes are full of ecstatic delight and, at the same time, absolutely sober. Sometimes he holds a feather and is smudging us with sage. He performs these age-old rituals with humility and, at the same time, absolute majesty.

* * *

I came away from the Southwest feeling: *"Now I know what it is!"* That primal merging with others, merging with sound, mindless, brainless merging—yet with it the spark of recognition, of focus, of awareness. For ten days we had lived in a sacred circle. Leaving it brought almost unbearable pain, a suffering I have come to expect whenever I enter a state much larger than ordinary consciousness and then must shrink myself again. The inflation of the higher state is one the ego does not readily release, perhaps knowing the beating it is going to take on re-entry. I learned my own unique way to resume ordinary life.

When the negative animus attacks come, I appeal to my female ancestors and feel shielded by them. Their help is always there. Ritual, of which our western culture is largely ignorant, helps us negotiate when relating to larger energies, if we are conscious enough.

Later, with the Colorado shamanic study group, we were to learn that you can't simply enjoy the ecstasy of spiritual experience without respecting the enormous power of them. There can be a lethal backlash if you do. I don't think that as a group we ever learned a safe way to handle these energies.

* * *

Three of us flew back to San Francisco the next morning. A graduate school friend, Harold, met us at the airport. "You look great!" he exclaimed.

I remember feeling very tired, but also changed. I was ready to leave behind my many years of analysis with Benjamin, which focused exclusively on what we Jungians call personal complexes, or my conditioned reactions to life. I knew that I could now contain many of the energies that previously had been dissipated in that work.

New Mexico's landscape was etched into my psyche. I understood that the "garment of brightness" from the Tewa song was being woven for me, and that, in time, perhaps I could "walk fittingly" on this earth. I had at last begun to put my earlier experiences of "non-ordinary" reality into perspective and to comprehend them.

With certainty in my bones, I was on a new path: To maintain a relationship with the serpent. To sing the song in my throat. To live a life that supports this.

Eight

April

We use a biodynamic spray, BD 501, on the lavender in April. We spray three times on each of the moon sign flower (air) days, when the moon is in Gemini, Libra, and Aquarius. I start stirring half an hour before sunrise, and because we spray only an acre and a half, I stir the water and the prep in the 12-gallon crock we keep up at our house. I face the east, staring out toward the ancient oak savannah meadow, and, as instructed by B.B., focus on my intention for our lavender and our land. While I am stirring, I sometimes play a CD of chanting or other music that moves me. B.B. said that vines like the Moody Blues, but I do not believe this is true, certainly not for lavender.

After I have been stirring for half an hour and before the sun has washed the top of the mountain to the west in pale red, Natalio arrives. I tease him by saying he needs to learn the stirring song. The "song" is actually a Hindu chant used in the fire ceremony that bypasses the knowing/unknowing paradox, taking the listener/chanter straight into the resonance of wholeness. Natalio looks at me as if I am nuts. His questioning look belies his conviction that I am a *bruja*. He laughs, a little nervously.

The BD 501 is also called horned quartz. If Natalio knew how it was prepared, he would be convinced that it was some form of witchcraft! One grinds quartz crystal into a fine powder and fills the horn of a cow (not a bull), then buries the horn in the earth throughout the spring and summer. The horn is then dug up and the quartz is stirred into pond or sun-soaked water, one-half teaspoon per five gallons per acre, for a full "German hour," just as Steiner suggested. We throw the water into chaos, over and over, so it can receive the divinity of the quartz. This solution is then sprayed into the air above the plants, to assist the assimilation of light and the solar forces. We follow a regimen developed by Maria Thun, an important biodynamic researcher, spraying the lavender in the spring, just before the flowers form.

* * *

Once I was invited to speak at a local food and wine center about growing lavender biodynamically. Everything went fine until I mentioned the moon signs and that it is best to plant, weed, spray, and harvest the lavender on flower days. Although these techniques are difficult to accept from a "scientific method" point of view, Thun's research on gardening in harmony with the moon signs shows increased yields and crops that keep longer. The basic system is to use flower or air signs with flowers; fruit signs with fruit; root or earth signs with root vegetables; and leaf or water signs with leafy greens.

I remember standing in front of the 30 or so people in my audience, watching several of their faces harden. The members of my family who were in attendance instinctively responded. My son Casey and his girlfriend Melissa, sitting in the back, were alarmed at the growing hostility they sensed in the mature ladies seated around them. My husband dashed to my side at the front of the room, ostensibly to hold a chart the wind was trying to dispose of. I blithely continued, describing Goethe's wholistic approach to the living world and its contrast to our more recent, so-called scientific way. I spoke of how this Goethean way might help us in farming.

Looking back I now see I should have foreseen that those practical gardening people would not be interested in the strange ideas of a 19th-century philosopher. At the end of my talk, I was shocked by the hostile reactions I received. "But it works!" I said as one man huffed off at the end. "The proof of the pudding is in the eating!"

* * *

Receptivity to working with natural forces came by my father. Despite all the Christian overlay of Midwestern farming communities, he was a water witcher, dousing for wells and for tiles that drained the fields and teaching me to do the same. As he saw it, dousing was a skill, like using a hoe or driving a tractor. Some of us could master it, others could not.

Country Christian is different from city Christian. Our farming fathers considered themselves to be stewards of the earth. From as early as I can remember, our local ministers gave talks that placed my family and all farmers in the greater scheme of things, cultivating a reverence for the Earth.

In our churches, the virtues of trusting and obeying were often expounded upon. "Consider the lilies of the field," the minister would read from the scripture. "If he cares about any one of these, why would he care any less about you?" My father and the other farmers were reassured by these words. Although they could do nothing about the unpredictable elements of rain and snow, heat and cold, God would provide for them in the end.

Yet these very farmers were led unquestioningly into taking advantage of modern science by using up the surplus ammonium nitrate from weapons manufactured during the Second World War, spreading it on their fields as a supposedly superior source of nitrogen. After all, was this not a way to pound "swords into ploughshares"? During the Nixon administration, farmers were encouraged to cultivate all the land, getting rid of the hedgerows, the pastures, the milking cows, the chickens. The small farm was seen as an inefficient and economically unfeasible way to produce food for the world.

Holding by holding, the small farms were consolidated into large farms which then used vast amounts of chemical fertilizers, herbicides, and pesticides on the crops. This turned out to be sociologically as well as environmentally foolish. For every six small farms consolidated, a business in the nearby town closed. This was the end of the way of life in which I was raised. Only now are we beginning to recognize the very serious consequences these trends triggered for our food supply, our health, our communities, and our planet.

* * *

Steiner's biodynamic lectures of 1924 were a response to a disturbing trend in Germany. Even then, chemicals were being used to increase production at the expense of good farming practices, and farmers were noticing that seeds were not keeping like they had in the past. Steiner attributed this to mechanistic science's belief that an element, like nitrogen, was the same if it came from composting materials or from a laboratory. He stated that the life force of the plant and the planet were at stake if we did not honor the living forces of the natural world, a relationship requiring humility and a receptive consciousness.

Later, in my psychology studies, I would recognize this same connection of ourselves to the larger whole in Jung's description of the ego and the Self. Farming humbles one into it.

* * *

Wearing a backpack sprayer, Natalio does the spraying of the preps for the lavender. We pour the 501 water into the two-gallon tank and lift it onto his back. He takes off, pumping a fine spray out and up. Rainbows arch over the lavender like blessings. Within weeks we will be surrounded by a sea of purple.

Nine

Trouble

Returning to Benjamin's office after the trip to the Southwest, I happily entered the combination for the padlock on the door. The lock clicked, and I entered the darkened hallway and walked past Benjamin's door to the waiting room. As I entered, I switched on the light to let him know I was there. In a few minutes he would open the door only enough to let me see it was he, and I would follow him into his office.

Benjamin's room was dark. The room's only window opened into a light well and sun cut across one corner of the room. Diagonal from the window was the patient's chair, which Benjamin frowned on me using. His chair was in front of the window. A flat beige couch with one end slightly raised stretched from his foot stool to the end of the second corner of the room. Above it hung an ancient carpet from the Far East.

This day I sat in the chair, ready to recount the experiences of the last month. He glanced toward the couch, and when I ignored the hint, he increasingly let his eyes rest in that direction. I knew that reclining on the couch opened me to feeling states less available than when sitting face-to-face with Benjamin. During most of my analysis, lying on that couch had been containing, allowing me to slip out of social convention and into a freer space where I could name whatever came to mind, undistracted by his facial expressions, or lack thereof, or, like today, any nonverbal cues.

Benjamin got the couch two or three years into our work. He warned me a month in advance that the couch was being made and would soon arrive. I warned him that I would not be using it, that it felt too intimate for me to lie down during our sessions. He continually encouraged me to use the couch, and several years into this struggle I relented, sinking into its expensive mattress as an exhausted person sinks into bed. For several years now I had entered the room and immediately lain down, relieved to be held by Benjamin's kind, soft voice and the firmness of the couch, both supporting me as my psyche was disentangled and rewoven under the pattern of that ancient carpet.

But on this day I sat in the chair. I wanted to tell him about the trip: the visions and the flute, the dreams, the drummings. I wanted to explain how I had come to feel that I could hold *myself* through some very difficult feelings, that I could be my own refuge now.

However, we were not conversing about these things. I gradually realized that the only issue alive in this session was whether or not I was going to submit to his wishes and lie down on the couch. Only if I did would he be available to listen. When I finally relented, the couch felt more hard than firm, and the darkness in the corner of the room enveloped me. I was quiet.

"Can you say what is going on?" he asked.

I suddenly felt like I had back at the beginning of our work. I had no answer. I was lying down on his terms, not mine, and I did not yet have it in me to say, "I don't want to use the couch today, I want to talk to you *directly*, person to person." Or, more dangerously, "I am healed. I do not need you in the same way anymore." The latter, I realized, wasn't quite true. I did still need his attention, as a new sprout needs sunlight.

That day, clouds were everywhere. I did recount the events of the trip, the dreams, the insights, but there was distance between us. I knew from experience the disquiet I would feel upon leaving the session if I did not move through the distance. Lying there, I soothed myself by remembering that it was always this way, that it took a while to get back together after a break. I remembered earlier times when Benjamin's vacations had felt intolerable, and I mentioned out loud that this had changed. But that day, nothing bridged the gap.

The winter of my first year with Benjamin, I had come across a children's book written by a nine-year-old child (I have always been sorry that I did not buy it and now can no longer find it). In the story, the boy finds a wounded pelican, nurses it back to health, and then is presented with a dilemma. If he keeps the bird any longer, it will become domesticated and never be able to return to the wild. He decides to release the pelican, and he never sees it again. I was overwhelmed with grief at what the child in the book sacrificed.

Driving home after my dissatisfying session with Benjamin, I remembered the story in detail. The lure of an illusory aspect of analysis, that one can find the perfect parent in the analyst, is hard to resist and important not to in the early and middle working phases. A kinship libido steps in, one that binds, and the transference intensifies in a way that easily can be confused with falling in love. Part of this intensity is the hope that old wounds will be healed this time, that the analyst will be a better mother or father.

* * *

After the trip to the Southwest, the full-blown intensity of my work with Benjamin decreased. My interest turned to dreams and imagery, to my own process in listening to what they might reflect about the psyche. I felt a difference with Benjamin that I had not felt before. I became less comfortable talking about my consultations with Don or the work with the group of people I had joined who were studying the overlap of analytical psychology and shamanism.

In later years, Don wrote a paper about one's relationship to therapists and spiritual teachers entitled, "The Power of the Transference."[17] In it, he describes and elaborates on Jung's outline of a three-tiered system of individuation, the process of moving toward consciousness and wholeness. The *first tier* involves the parental transference and shadow material. My first eleven years with Benjamin reflected the work of this tier, in which relationships are often a focus. The patterns of feelings that we developed as children toward our intimate others are typically transferred onto the analyst, and analysis works to make these patterns more conscious, in part via their presence in the relationship with the

---

[17] Sandner 2006. pp.15-27.

analyst. Often the lived experience with the analyst changes these patterns. During this phase, therapeutic boundaries have to be firm to hold the integrity of the therapy, the way incest taboos protect families.

Goethe, and then Heisenberg, brought a shocking idea to science: there is no such thing as an impartial observer; the act of observation changes what is being observed. Jung applied this idea to analytical psychology. In analysis, this "observer" is often first the analyst, and then, over time, the strengthened ego of the patient. The so-called observing ego of the patient is trained into a kind of self observation of movements in the unconscious, and in time the analyst is no longer needed to perform this role. Achieving this "observer" status gives the patient a leg up on old patterns, and the relationship with the analyst moves from the first into the second tier during this process.

The *second tier* involves, in Don's words, "establishing the relative reality of the inner and outer world."[18] In this stage, done most often in mid-life, the work is with the contrasexual aspects of the psyche, the *anima* or *animus*, and withdrawing one's projection of these energies from outer figures such as one's partner or analyst.

The *animus* for a woman (in the classical Jungian sense) is the masculine aspect of her psyche, such as the left-brain functions of analyzing and categorizing, seeing things in a linear fashion. Often the *animus* is projected onto men. In classical Jungian theory, women are naturally much more based in the "feminine," such as right-brain functions, concerned with wholeness, relationships, and beauty.

For men, the *anima* carries these right-brain functions, and "she" is projected onto outer women, particularly a partner or a would be partner. It is through the *conscious* connection to the anima/animus that we strengthen our connection to the Self, which includes the ego complex but is also so much more.

The work of the second tier prepares us for the *third tier*, in which "the unconscious takes on a reality of its own... reality as solidly present

---

[18] Sandner 2006, p.21.

as anything in the external world."[19] Don describes this as an initiation on the spiritual path, most often a task of later life, but a course that shamans or spiritual acolytes take much earlier. Often the object of transference in this tier is someone other than the analyst, and the terms of the transference relationship are often different than those in "analysis proper." "The analysand here must be the seeker, who is no longer a patient, and must take full responsibility for his own salvation," Don claims. "The analyst is no longer a doctor or a guru but more of a teacher-mentor with commitments and obligations worked out between the two of them. ... This has resemblances to all the spiritual and shamanic paths in the world with one crucial difference. It is to be done in the light of greater consciousness."[20]

The following year brought increasing difficulty with Benjamin as we shifted "tiers." It was quite painful for me and I am certain it was for him, too. One day he told me that for a while he had thought we were on opposite poles of the continuum that was a source of a great deal of conflict in our larger Institute, me thinking that the so-called "clinical" and "developmental" approaches that he championed to be wrong, the "archetypal," right. I answered that I had been feeling that he saw *me* as wrong if what I said was different from the way he thought or felt.

I was afraid to say to him, "I'm worried that you'll work regressively with material that is not only regressive, but is something else as well. I am worried that I will *never* get out of analysis." I did not know how to trust myself on this. Was this a complex? Was I again meeting some place in myself that just couldn't trust? Should I abdicate, trusting he had the larger, correct perspective, as I had for these many years?

I questioned my thrice-a-week appointment schedule, which was both a financial and logistical hardship. He said, "You need to be here three times a week for a certain deep level of work. Whether or not you need to do that level of work is another question." He added, "You want

---

[19] Sandner 2006, p. 22.

[20] Sandner 2006, p. 23.

the Cadillac version of analysis. To go the whole way." And so I did not cut back for another few years. The grief was profound when I finally did, but I discovered it was something I could bear.

Mary Jo Spencer, one of my two training analysts at this time, felt that a depth transference might never get resolved. "Transference is upsetting," she said, "because it is against keeping things unconscious. It allows the alchemical *nigredo*, or darkening, to become evident, and it happens outside of ego consciousness."

For the first time I wondered about Benjamin's own infantile states and his love of working at that level. I remembered a dream I had before the trip to the Southwest in which another candidate on the trip called to me in warning not to go upstairs. "Be careful of you and the baby's dependence," she had said. At the time, Benjamin interpreted this as a warning that the relationship with Don was threatening the analytic work and my fledgling self. I had accepted this as true.

But now I wondered if in order to move to another tier, I had to confront the infantile dependence that was set up between Benjamin and me. Was Benjamin inadvertently encouraging me to stay at the first tier? I questioned him on this. Would he be able to help me terminate my analysis, particularly when he had not terminated his own (which I knew)? In classic form, he just answered my question with a question, "What is it like to be with an analyst who has not terminated his own analysis?" My work with Benjamin was increasingly scary. I worried. Did I have it in me to leave by myself? Would Benjamin hold me back, even if I was ready?

"We need Jung as a holder of this container," I wrote in my journal at the time, "because he had sensibilities and awarenesses we don't yet, and just reading him opens something. I also feel this with Don Sandner. It is not my experience of Benjamin. He is more of a mother, and yet the work with him has healed me. It is what all the rest rests upon... this work with the subtle body."

I was heading into a whole new territory, but I barely knew that at the time. As healing as my work with Benjamin had been, I was going to need a very different kind of guidance from what he had provided.

Ten

Norma T.

Norma T., whom I have come to consider my spiritual mother, has been dead for several years now. When she passed, her assistant James told me that she would be able to teach more from the other side. I thought at the time that he was full of it, but I was wrong. Now I understand that, just as she taught, the larger part of us is not in our bodies at all, but in the spirit world. When we go, we are completely in the spirit world. We're on a *sliding scale* of incarnation, with death at number 10; death has nothing to do with one's ability to communicate.

Norma T. gave me a perspective that opened into the world of spirit, which I believe is the bedrock of sustainable farming and the soul work at the core of Jungian psychology.

I first met her in 1992, five years before the one and only séance I ever attended, and after a disturbing experience with a threatening phone call. I was a third-year candidate in my analytic training program, living alone with my sons, aged 9 and 11. Lynn had been gone two years and I still felt vulnerable. The man who called knew my unlisted phone number and address. He said he had been watching me and would hurt my sons if I did not do as he said.

I hung up immediately, not waiting to hear what he had in mind. I called friends; I called the police; I bought and learned to use a stun gun. A short while later my home was broken into and my office phone number was charged with many long-distance calls from other locations. Although the incidents may not have been related, the cluster of events were alarming. I got a black poodle, big enough to protect us and gentle enough not to maul my sons. A friend suggested that I contact Norma T., a psychic he knew quite well, to investigate the frightening events on another level. Skeptically, I called for an appointment.

Norma's office was in an old Victorian on Van Ness Avenue in San Francisco. The long stairs to her second-story office were rickety, and as I climbed them for the first time I wondered if I had gone out of my mind. I knew others would think I had—namely Benjamin who did not approve of any such consultations outside of the alchemical alembic of the therapy.

At the upper landing, I noticed that the door on the left, into what I later learned was the hallway to Norma's office, was blocked off. Straight ahead was the door into the reception area, which I soon found was full of metaphysical books for sale, as well as tapes and CDs, sage wands, tarot cards, crystals, jewelry, and James.

James was Norma's secretary and, I came to realize, her protector. Like Don, he too had the shaman's wandering eye that kept track of those who entered and left, including their energetic states. He attended to almost everything Norma needed, as well as doing all the things secretaries typically do: collecting fees, making appointments, answering the phone. On that first day, he told me Norma would be right with me. I waited, getting more anxious by the moment, and disguised my pacing by browsing through the merchandise.

"Right with me" stretched to forever by the time Norma came to the door of the reception room, greeting me with warm professionalism. She was of considerable stature, her graying blonde hair pulled back in a tight bun. She wore a long, flowing skirt and a hip-length purple vest over a turtle neck sweater to stave off the cold of the building, a problem during every season of the year.

Ushering me back through the darkened hallway, she led the way into her office, which had once been the living room of the reconfigured flat. Faded floor-length olive-green drapes were drawn over the large windows that faced the street. A massage table blocked the front of an ornate fireplace. In the middle of the room stood a large desk with a well-used, high-backed antique chair on either side. To the right of the door was a small black velvet tent just large enough to fit over a person sitting on a straight-backed chair. Behind the tent was a bookshelf with ceramic owls perched on the top. In fact, owls sat on every ledge of the room. This was definitely a pagan lair.

"Have a seat," Norma said, gesturing toward the chair in front of the desk. As she clicked a tape into the recorder, she began casually chatting with me, yet I came to know later that she used this opportunity to examine my aura. I squirmed around in the broken-down chair, trying to find a comfortable position.

The desk was covered with the tools of her trade: a large hourglass, a round crystal the size of a honeydew melon, a pendulum anniversary clock, her tape recorder, a calendar, a six-inch diameter magnifying glass, teapot, cups, books, papers, more papers, and in the cleared area in the middle, a worn stack of tarot cards. She obviously did not subscribe to the de-cluttering practices of *feng shui*. Yet, although the room faced one of the busiest streets in San Francisco, it gave one a sense of being apart from the bustle. Buses groaning at the corner nearby soon became white noise.

As any good farmer's daughter would, I waited politely to be asked why I was there. It soon became obvious, though, that we were not simply going to ask and answer each other's questions. "Your aura is so close to your body that you are going to be sick if you don't rest," she said abruptly. I stared at her in shock. I knew I was tired but didn't realize it was a dire situation. "Your second chakra is kind of muddy and is moving very slowly." I had no idea what she was talking about.

After this warm-up, she pushed the tarot cards toward me and told me to shuffle them. As I did so, we continued talking. She asked about my work. I told her I was a psychotherapist and a writer. When I spoke of my most recent article, she told me my aura flared out, a good sign. She instructed me to cut the tarot pack three times with my left hand, concentrating on the question that I would like addressed, then to spread the cards into a fan and choose seven of them, which she placed face down in a line.

As we talked, my life story began to emerge. I told her about my divorce, the analytic training program I was in, and Donald, the man I had recently met whom I had been seeing for a couple of months.

"Is this serious?" she perked up, looking interested.

I was not even ready to entertain the idea of serious, and told her so. "Besides, he is quite a bit older," I added.

"Don't be an ageist," she scolded. "This could work. You are of the same frequency. It may be only past life memory, but it is possible that you will be together in this life, too. He would be a good catch."

Silenced, I watched as Norma reached to turn over the first card. I don't remember which card it was but I vividly recall her nonchalant comment: "Your grandmother is standing there on your left." I felt a chill go up my back. "She is a protector," she added. "When did she pass?"

"I was thirteen," I said. "She was like a mother to me." Suddenly I knew this was why I had come: for some verification of the objective reality of the spirit world. I was surprised to feel my throat close, as if I were about to cry, and I struggled to stifle my tears. My siblings and I had been grieving my mother's mother since she was carried on a stretcher down our front walk for the last time.

"Oh my," Norma said, in a way that I will always remember, full of compassion for my emotions, but not indulgence. I am sure that, being a medium, she was used to people's grief, as well as their lack of awareness of their continuing relationships with the dead. "Now what is it that you would like to ask?"

I told her about the phone call, the threat, the break-in, and the long distance charges.

"Did you tell this fellow you're seeing about the phone call?" she asked.

"Yes," I said. "He offered to come stay at the house."

This seemed to satisfy her. "He would be a good catch," she repeated. "He cares about you. And there is someone else out there watching him. He has a lot going for him." She softly suggested I tell him that I liked what was developing between us, yet continue taking it at a slow pace.

"Every time I speak of moving ahead with this relationship, your heart contracts. That will change at the beginning of the year. Then the relationship will go one way or another."

Norma told me that she was not too worried that the caller would do anything more, but she did suggest some protective action: boiling cut-up ginger in water, adding lemon juice, and using this concoction to wash all the window sills of the house and make the sign of the cross over each door. She said to place statues of St. Michael on the window-sills, facing out. She also told me to stay away from the house for six weeks. Nothing minor to accomplish! She explained that the best way to avoid provoking the forces behind the caller was to stay away.

So this was our beginning. In time, Norma would reinforce and name the "organs" of my intuition, as her grandmother had done for her, which help differentiate perception, imagination, and fantasy. But that would come later. It would be several years before I was ready for my real work with Norma T.

Eleven

Colorado Group
First Three Years

In the spring of 1992, I attended my first symposium on shamanism and analytical psychology. Several such symposiums were organized by a small committee comprised of analysts and people with a special interest in shamanism, including Don Sandner. Because we met most years at Woodspur, an old ski lodge in Winter Park, Colorado, we became known as the Colorado Group. There were seven symposiums in all and I attended the middle five.

The structure was unique in that all participants were equally responsible for the presentations. Attendance was limited to 30 participants, most of whom attended every year. When someone new wanted to join, he or she went on a waiting list until someone resigned. We split the basic expenses of food and lodging equally. Presenters generally were not paid. Steve Wong, one of the organizers, recorded the sessions, and later he and Don published the edited papers in a book, *Sacred Heritage: The Influence of Shamanism on Analytical Psychology*.

Initially, the Winter Park symposiums felt to me like a natural extension of the trip to the Southwest. Participants included Jungian analysts and candidates, Native American medicine people, and others from various walks of life, providing both diversity in background and common assumptions about the role of the healer and the terms of the therapeutic relationship.

The symposiums were always held in mid-May. That first year I flew to Denver with a couple of candidates who had been on the earlier trip. Steve arranged for local participants to collect us at the airport and drive us to Winter Park, an hour-and-a-half trip. Winter Park is at about 9,000 feet, and in May there is often still a fair amount of snow. By the time we arrived at the lodge, I had a headache from the altitude and the excitement. I had come to expect headaches when there was so much energy aroused. We all complained of being headachy and drank huge amounts of water.

After getting our room assignments, we gathered in the central hall for dinner, always served buffet style. Then gradually we trickled into the main room, an expansive space with a wall of windows overlooking the highest of the Rocky Mountains. Pulling couches into rows, we each found our own comfortable space for the evening lecture.

Don Sandner gave the introductory talk. He spoke of Jung's view of shamanism and discussed the difference in dynamics between analytical psychology and shamanism. "Shamanism involves a nonintellectual, direct knowing that affords direct contact with the unconscious," he began. "If this way of knowing comes into analytic therapy, it cannot come in the way we usually practice therapy... Central to shamanism is the cosmic tree which connects the lower, middle (where we reside) and upper worlds. To travel between these three levels of reality necessitates an opening—perhaps a pond [like the pond of my childhood dream, I noted], a tree, or woods. Then we are transported from geographic to mythical reality."

Don interwove Navaho mythology into his discussion. Although the Navaho are not a shamanic people, their mythology involves the formation of a body that can navigate in the spirit world. For the Navaho, the aim is to follow what is called the pollen path. Pollen represents the *essence* of life. If we live life completely, they believe, we will leave no evil ghosts behind. "Essence" *is* the subtle body, called the diamond body in other traditions. It is only through living in the physical body that we build up essence, the subtle body. An unrealized life—in which some parts are repressed—results in neurosis and, according to Navajo tradition, ghosts after death. When we die, Coyote will eat our bodies so that our spirits can be released and go on.

Don explained that the shaman is the technician in this world of subtle bodies and energies. He can leave his physical body to do feats for his community: look for game, retrieve souls. He does this only after developing the spiritual organs of the subtle body. His initiation involves dying to the old world of the five senses, then acquiring spiritual organs which allow for extraordinary "sensing." This is all part of the building of "essence."

In the open-beamed grand room of Woodspur Lodge, thirty of us scribbled notes. The backdrop of moonlit white peaks behind Don punctuated the rarified environment. I wanted for nothing that evening or in the days that followed. Clearly I was exactly I needed to be and, by some act of grace, I was there with the right companions.

The unique thing about the Winter Park symposiums was the format. There were lectures, like the one Don gave the first night, interspersed with rituals led by Pansy, a Lakota medicine woman, particularly the *inipi*, or sweat lodge ceremony. On Friday morning, we were divided into groups to begin the building of the sweat lodge in the parking area of Woodspur. One group gathered the rocks and one cut saplings. I cut saplings.

Gathering young saplings to bend into the curved shape of the lodge, representing the rib cage of the Mother, was no easy feat, especially when the snow is three feet deep. We slogged through drifts, offering, in the native tradition, tobacco and a prayer of thanksgiving to each tree before cutting it. The young trees had to be thin enough to bend and long enough to arc over so they could be lashed together. We bent each tree slowly, pressing a knee to the inside of the trunk as two of us inched the tree toward the center of the circle. We cut two trees for each of the four directions of the wind and for each of the diagonals, and then more to encircle the structure in two layers. We tied the bent trees into the curved form with colored cloth representing the six directions: yellow for the east, white for the south, black for the west, red for the north, blue for the sky, and green for the earth.

The door of the lodge faced west, because spirit travels in through the west door. In the center of the *inipi* we dug a hole for the hot rocks, piling the dirt from the hole directly west of the door, halfway to the fire pit. This became the altar, where Pansy placed her staff and pipe and we participants placed personal objects to be purified.

Pansy always built an enormous fire to heat the twenty-eight rocks embedded in it. The group collecting these rocks also offered tobacco and a prayer of thanksgiving as they worked, taking care to select rocks that would not explode in the heat. Once the fire was lit, someone tended it constantly: the ceremony had begun.

We sat watching the fire burn to ashes as the rocks heated, while Pansy taught us about the ceremony and the Lakota way of life. We sat as close to the fire as we dared to ward off the cold of predawn or the pelt of rain. Neither time nor weather determined when these ceremonies took place.

Through her direct teachings and her remarkably grounded presence, I was infused with a more conscious way of interacting with the energies of the trees, the rocks, the animal spirits, the ancestors. She was like a wise tortoise, slow moving but sure. Never did she step over an object. Out of reverence, she went around it. Never did she take any natural object or dig a hole without an offering of tobacco. Always she went about things with her own strong sense of pace and decorum.

In later years, a group of women from the symposium met with Pansy once or twice a year to learn the discipline of the lodge. Once she shocked us by leaving almost immediately after she flew in to San Francisco from South Dakota, saying it did not feel right to do the lodge at that time. This does not jibe with the American way of doing things!

In those meetings with Pansy, we women learned another sense of timing, sometimes finishing a ceremony, limp from the sweat, only to be instructed to immediately start the fire for another. We submitted to Pansy's teaching so we could learn a way determined by what spirit dictated.

* * *

This is how the ceremony proceeds: When the rocks are hot, Pansy tells us to prepare for the lodge. The women wear long cotton gowns that cover our knees, the men swimming trunks. We wear no jewelry or metal of any kind. Once we assemble, the rock carrier scoops glowing coals from the fire into a shovel and Pansy drops a handful of sage onto them. With the smoke this releases, she smudges each of us to cleanse our energy fields. Then we enter the lodge, the womb of our Earth Mother, humbly, barefoot and on our knees, crawling clockwise around the fire pit. The lodge is covered with blankets and black plastic, so it is very dark inside. When we are all seated, Pansy tells the rock carrier to bring in the hot rocks. Once, when only we women did the lodge, I performed this task myself. It is difficult. You coax a glowing red rock about the size of a large melon onto a pitchfork, and then you carry it carefully to the door, balancing it so that it does not roll. Pansy guides it to the pit and a helper sprinkles it with tiny pieces of cedar needles, saying, "*Mitakuye oyasin*" —"All my relatives." We are all connected.

Pansy then passes around a bundle of mountain sage sprigs she collected at the Sun Dance in August. We each take one and, in the growing heat, the sprigs release their pungency. I chew a piece of leaf, a taste that has become synonymous with the lifting of the veil for me, the veil into the other world. Sweat is already breaking. I feel streams of it running down my stomach. Only half the rocks are in; it will be a very hot lodge. I feel my anxiety rise and take another bite of sage. The familiar, bitter taste reminds me of other times when I survived the heat of the lodge. The earth beneath me is cool and this soothes me.

At last, all twenty-eight rocks rest in the carved bowl of earth in the center of our circle. The rock carrier enters and closes the door. Blankets are adjusted until we are encased in velvet blackness. The heat intensifies. There is the intimacy of Pansy's clear voice as she invites prayers. One by one we speak, each person's voice soft and distinct. We know each other by our prayers. Pansy sings "spirit calling" songs, all in the Lakota language. Finally she declares, "The ancestor spirits are here!" My skin erupts in goose bumps. The heat grows and the singing continues. It seems that we have been here forever.

Finally Pansy signals to the door keeper to open the door: "*Mitakuye oyasin!*" Cool air wafts into the lodge, a welcome relief from the heat. We can see each other in the dim light. Some of us are lying down, noses close to the earth. Others are sitting, faces covered with towels. Some are crying. Our faces are those of suffering: suffering the heat, suffering the personal pain we bring to the lodge for healing, suffering the pain of the person next to us. In this space our hearts have opened.

We welcome the fresh air. This is the first door.

All too soon Pansy instructs the door keeper to close the door and again we are in total darkness. She then ladles water onto the rocks and it spits and hisses. This is the second door, a time of intense praying. Not wanting to be burned, I cover my face with a towel. The heavy cotton flannel nightgown I wear is wet with sweat. A wave of nausea comes, triggered by heat and dehydration. I long for a sip from the ladle of water Pansy pours over the hot rocks. I focus my attention on the person next to me. When we can't stand the heat, that is what we are instructed to do. I pray silently for her. I pray for my family and I pray for myself. Pansy sings traditional songs in Lakota and beats a rhythm on her drum. The sound transports me into another space. The heat grows and the singing continues.

Visions come, materializing in the black before me. An enormous golden serpent stares at me in the most majestic way. I sit in awe, feeling it slowly enter my feet and slither up until its golden head rests on my heart. I hear its words, words which are not yet sound. In the darkness, I see the ancestors, blue against black. The snake teaches me how to listen to them, relaxing into a dialogue. My physical suffering fosters the non-distracted state of mind to be receptive to these visions. The snake points this out, too. Then: "*Mitakuye oyasin!*" The door opens, we are blessed with a breeze.

So I learned to expand my psychic experience in the structure of the *inipi* ceremony and under the direction of Pansy. To this day I remember the freshness of crawling out of the lodge, reborn, after the door opened for the fourth time. Soaked with sweat and steam, humbled, our hearts opened, each person's presence pure and authentic. Pansy hugged each of us as we emerged, and we hugged each other. I remember the clear, thin air—so precious; the colors of the trees—so vivid. Blessed life, such a gift.

* * *

In the main room of the lodge on the last evening of the first year's symposium, we were led in a psychodrama. We donned animal masks and danced. When the plumpness of the full moon rose above the mountains outside the windows, our leader simply stopped and, in unison, we all raised our arms to the moon, as if we were helping it rise. The level of involvement was so energized that our western egos were confused. We knew that we did not raise that moon, and yet our presence felt absolutely necessary to the event.

Jung recounts a visit he made to the Southwest, during which Taos Pueblo Ochwiay Biano told him that the Pueblo religion helps Father Sun go across the sky, that what the Pueblo Indians do in ritual, they do for us all. That if they stopped, within ten years the world would be dark. Jung describes Ochwiay Biano's life as being "cosmologically meaningful."[21]

He goes on to say: "Knowledge does not enrich us; it removes us more and more from the mythic world in which we were once at home by right of birth. If for a moment we put away all European rationalism and transport ourselves into the clear mountain air of that solitary plateau... if we also set aside our intimate knowledge of the world and exchange it for a horizon that seems immeasurable, and an ignorance of what lies beyond it, we will begin to achieve an inner comprehension of the Pueblo Indian's point of view."[22]

The Pueblo Indians were and still are as immersed in their world

---

[21] Jung 1965, p. 252.

[22] Jung 1965, p. 252.

view as we Western Europeans are in ours. The two views, generated from very different ways of being in the world, are part of an immense puzzle. That evening when the full moon rose over the Rockies, I experienced both perspectives.

* * *

The symposium of 1994, my third year as a participant, was one of the most exciting and certainly the most tragic. That year I gave my first professional paper, "Shamanic States in Our Lives and in Analytic Practice."[23] I did so in the spirit of the symposium, wanting to share in a professional context some of the intuitive and psychic ways that I approached my work as a psychotherapist. At the time, I had just been advanced into the control stage of my candidacy, in which candidates work with control/ training analysts on cases that we then write up for our final presentations to the certification committee.

In stark contrast to the difficulty of the early years in the candidacy, these middle years had been smooth and fruitful. I enjoyed the control work, and the seminars were interesting. I had taken a year off from the seminars to adjust to major changes in my personal life and was finishing the four years with a group that worked well together. Now that my energies were less tied up in the difficulties of a major life transition, I wanted to give something back. Writing and presenting the paper were steps in taking full responsibility for my own salvation.

The night before I gave my paper, I had three nightmares. The first two contained personal material reflecting my anxiety about presenting, but the third carried another level of anxiety. *In the dream, I am parked in what I think is a parking lot but is actually a driveway. I move my car inadvertently into the way of a giant driverless street sweeper that is cresting the hill. I realize, as it bears down upon me, that I am going to die! I am re-*ally *going to die!*

Awakening from this dream, I got up and walked up the mountain

---

[23] Sandner, Wong 1997. pp. 71-77.

to wrestle with its meaning. The morning was overcast and the air cold. I seriously considered not giving the paper. Was the dream suggesting that there would be dire results? I finally decided that this was another activity of a very negative animus. I was talking about something in public that I'd always kept private and, I thought, this dream reflected the equal and opposite reaction, the voice that wants to reestablish homeostasis.

Presenting the paper was a maturing experience, one I enjoyed very much. It felt absolutely right to voice my own experience. I also told the group my dream. Later, as we waited in line to enter the sweat lodge, Don commented on my paper and then on the dream, also attributing it to negative animus energy. I remember saying to him, "I hope so."

The next day, we all entered the sweat lodge as usual, on our knees. Some twenty of us were packed in together. Pansy called in the spirits and the drumming and songs began. At one point, I looked across the pit of hot stones and saw a white phosphorescent skeleton sitting in the circle next to one of the organizing participants, Poul. The image was there for a lingering moment, and then it wasn't. I can count on one hand the times that I have had visions of this kind. They are of an otherworldly character. The light is of its own source, especially clear. Then the image is gone as quickly as it came, like a switch has been flicked off.

Even now I am appalled at my reaction that afternoon, one that I have always regretted. I registered seeing the skeleton sitting there with its arms on its knees, like all the rest of us, but I did nothing about it. I told no one. It was an example of what Jung describes as a purely aesthetic attitude, which can be dangerous. This time it was catastrophic.

After the ceremony, the evening proceeded, energy high all around. Poul gave a presentation, his first, as well. We were all high on our experiences, both experiential and didactic, and grateful to be among people who shared our passion for this work. We were so excited that sleep almost seemed unnecessary.

The next morning, the last of the symposium, our closing circle seemed to only increase the intensity. I noticed that despite the euphoric atmosphere, I was seeing black dripping in that other realm, thinking the worse thoughts about these people with whom I also felt bonded. Again, I did not consciously attend to the image. It was as if there were a thick, dark, impenetrable veil in place that blocked my ability to have judgment about what I was perceiving.

Sunday afternoon I flew home. Donald, then my soon-to-be husband (I had taken Norma T.'s advice seriously and our relationship had flourished), picked me up and we drove to a restaurant for dinner. As we stopped at an intersection to turn right onto a highway, I remember watching the shadow of an approaching car merge with ours, just before feeling the bump! The words came: *You got off easy.* We had been rear ended, sustaining some damage to the car but not to us.

I took the next day off. I had learned that I needed integration time after the symposiums before I was ready to return to work. On Tuesday morning, I had an early appointment with Don Sandner. For some reason, I took him the sloughed-off skin of a large gopher snake I had found on our property. It was the only thing I ever gave Don, a remarkable thing considering what was to come.

Don met me in the waiting room and silently ushered me into his office. He looked exhausted, in stark contrast to his demeanor over the weekend. As we sat facing each other in the darkened space, he looked at me for a moment, then said, "Poul has been murdered." We sat together in shock. He explained what he had just learned, giving me the very few details that were available. After the symposium, Poul, still in the euphoric state that we were all in, dropped Don and others off at the airport, planning to spend a bit more time in Denver. Evidently, he picked up a hitch-hiker who then robbed and murdered him in a motel room. I wondered if this happened at about the same time that the car rear-ended Donald and me. I thought about the parking lot of my dream and the approach of the deadly street sweeper.

When we see into this other realm, we have a responsibility for what we see, as we have a responsibility to the images that appear in our dreams. Were I doing it again, I would have at least told Pansy, our spiritual leader during the time of the vision. Would this have saved

Poul's life? Possibly not. I will never know.

The dream took on powerful new meaning for me. Death *is* a driverless street sweeper, cleaning us off the road. In our culture, death is something we typically try to ignore. We want to prolong life, steal years from death. But in the end we are all swept up. We shed the skin of the physical body and are put to the test: How large is our subtle body? How strong is our essence? Large enough and strong enough that our consciousness can survive death, this time?

Had I participated in Poul's death by not speaking of what I saw? Steve was upset with me that I had not acted; Don told me that there was nothing I could have done, and when I kept bringing it up he told me to stop thinking of it. Maybe in terms of averting Poul's death, he was correct. But to this day I wonder: Had Poul been less high from the energy of the weekend, less open in absolute trust, would he have been more cautious? Was my vision of the black, dripping substance inviting me to attend to darker feelings which would ground myself and others? Was this another warning I had disregarded?

I will never know the answers; I have to be content to articulate the questions. As I sat across from Don the day he told me in exasperation to stop thinking of the skeleton, I knew that this was not the guidance I needed, that I could not rely on Don's mentorship for direction in this matter.

This was the beginning of a painful process of taking back what I had projected. Don was finally assuming human proportions, as Benjamin had before him.

Twelve

Differentiation

Don was not the only person unable to help me understand my apprehension of the skeleton. Benjamin made it very clear that he was not interested in entering this territory. "You are much more enamored with these experiences than I," he said.

I stared at him in disbelief. "What do you mean, *enamored?*"

"Thrilled," he elaborated.

"I *am* thrilled to be able to have access to part of myself that apprehends in this way," I countered.

He explained that, although he was like me in this way, intuiting a patient's dream before a session, he was not as *enamored* as I by the knowledge.

Was he suggesting that I was inflated? I felt hurt and confused. "Don't you know the meaning to me, Benjamin? I celebrate being able to own the part of me that knows in this way. Your word 'enamored' devalues my experience. I need structure to understand what is happening.

"This is the source of my lack of confidence," I continued, "and I can tell almost no one but you! What belongs to which realm has been confused, and I need someone to help me sort it out. If I reside at one pole or the other, I am inflated with too many energies of the Self, or weakened with a limp ego. I need to recognize things of the spirit for what they are!"

At the end of the session, he apologized. But this was just the first round of many.

* * *

That summer Donald and I were married and my sons and I moved to Donald's ranch in the Napa Valley. Each day Donald and I took the dogs on the trails into remote mountainous areas that reverberated with birdsong and silence. I felt different, secure and loved by Donald. Something inside me was pushing out. The long years of my analytic work had strengthened me enough that I now had access to aspects of the psyche I was struggling to understand. It was becoming increasingly clear that I needed a different kind support and guidance in my analytic work. And I was tiring of my drive to my offices and twice weekly analysis at widely dispersed locations in the Bay Area.

Late in the fourth year of my candidacy, I made a second visit to Norma T., who was alarmed at my exhaustion. "You are putting so much energy into blocking your second chakra," she said. "It is running like molasses, muddy and brown. Your health aura is only three inches from your body. It should be radiating like an Indian head dress."

She told me that my strength was intuition and, beyond intuition, mysticism. If I could get empowered by rest, laying off my mind for a while, I would be in touch with resources never imagined. She also told me that I needed to quit going to see Benjamin twice a week and remarked that my plan to cut down within six months was not soon enough.

Benjamin was appalled that I would visit a psychic. "You don't need someone to tell you about yourself," he said.

"I don't go to psychics to *know* about myself," I corrected him. "I already *know*. I go to reinforce the field that brings my knowing, that field that can be so impoverished. I know it with you in following the paths that images make. But when we get to this issue of spirit, something else happens. I hit a wall, made in part by your attitude."

Benjamin's reaction confused me. I knew that he was creative, with his own artistic discipline. I had not thought of us viewing reality that differently. Yet I felt compelled to understand this other way of perceiving reality, and he did not.

As I became more aware of an inner spiritual reality, I also felt its objectivity. Not everything in this "inner space" belonged to my own psyche, yet Benjamin wanted to clothe it in psychological language. I felt an increasing hesitance to speak about my experience for fear of criticism or correction.

In his insightful foreword to Gerhard Wehr's *Jung and Steiner: The Birth of a New Psychology*,[24] Robert Sardello addresses the differences in Jung and Steiner in their considerations of "archetypal realities." He writes that Jung addresses archetypal patterns and images as contents of soul life whereas Steiner sees "archetypal beings" as shaping the human being and the earth. "In Steiner," Sardello says, "image is activity, the pure activity of forming or coming into form of the *actual presence* [italics mine] of spiritual beings. Image is the first way in which we can be present to the activity of spiritual beings. Image [in Steiner]… is a decidedly spiritual notion, while in Jung it is the very heart of soul." [25] This difference in emphasis between psychological reality (Jung) and spiritual reality (Steiner) is where things get controversial.

In psychological development, the ability to symbolize is paramount in the development of soul. Symbolic work with an image is the mysterious process of seeking the essence of an image and understanding its subjective impact upon oneself, as meaning. Jung lamented that modern man is in deep need of the meaning symbols offer through their resonance with the unconscious. For example, in my monastery dream shortly after my admission to the training program at the Institute, I experienced the importance of the animus (in the dream, Don Sandner) as psychopomp in the work of the inner (symbolic) monastery. In waking life Don's spiritual development made him a good model as a dream figure initiating balance between the inner and outer worlds. Like the shaman, he could move between the worlds of the seen and the unseen,

---

[24] Sardello 1990.

[25] Sardello 1990, p. 16.

outer and inner, psychological and spiritual.

The monks, on the other hand, are residents of the monastery. The monastic experience is known to be one that often includes psychological descent. "Monastery" has its roots in ecclesiastical Greek *monasterion,* or *monazein,* meaning *"live alone."* Although I was joining a community of the Institute, the journey itself would be a solitary one of much psychological suffering. The development of the spiritual animus would become critical to restoring balance between the inner and outer worlds. Through working symbolically with these images I differentiated and developed aspects of myself that Jungian analyst Edward Edinger might call soul, receiving guidance from deep aspects of my own psyche as to the proper course.

Edinger calls symbols "transformers" of psychic energy, bringing larger energies into the development of soul, an important aspect of incarnating spirit into matter.[26] According to Jung, developing the capacity to work with symbolic images in dreams or active imagination is vital to the process of individuation. To concretize an image that belongs in these symbolic realms as an "outer spirit being," as can so easily be done in these "New Age" times, short-circuits this transformer function of the image as symbol. In the case of the dream, this might mean taking the monks to be actual spirit beings. While I do not exclude this possibility, had I overlooked their symbolic aspects, I would not have appreciated the need for the strengthening through psychological descent. Furthermore, besides a strengthening of those inner monks dedicated to the rigors of in the monastic life, a new orientation or new birth of the spiritual animus was being called for, as reflected in the baptism.

All of this said, to label an image as a symbol when it may represent *a spirit being* poses a problem with more implications than many psychologists understand. Psychologically, it may appear to be the safe thing to do. No one is going to call you psychotic for it! Besides, we Jungian analysts are well acquainted with images reflecting the development of soul and know what to do with them. We spend years in our

---

[26] Edinger 1974, p. 110.

own analyses and training to learn to do this.

But different processes are required in dealing with imagery representing the activity of spirit beings, and most of us are not trained in these processes. Had I been able to accept my apprehension of the skeleton sitting next to Poul as a direct perception of spirit energy present in the sweat lodge, I might have averted the acting out of this energy through naming it.

Both aspects of image can be present: that which is objective and *not* us, and that which has deep resonance to larger aspects of the personal psyche. While the inner monks had meaning to me in a symbolic sense, I will not rule out that they may also be spirit beings in their own right concerned with the psychological and spiritual balance in my professional organization. Differentiated and yet held all at once, both ways of perceiving can be like the harmonic notes of a chord.

* * *

Benjamin and I argued often.

"I am in an egg stage," I told Benjamin once, "and usually you know how to incubate. But now you crack the egg open and tell me what should be there. This makes you unsafe!"

He countered, "You have been wary of me from the beginning. In terminating we need to do another piece of work."

"What work is that?" I asked.

"I can't say," he replied. "My intuition tells me that there is another piece of work to be done."

In another session, he worried that I was externalizing out of a desire to accept the possibility of spirit objects. Again we argued. Later, he revised this. "To an introvert," he said, "the outer sometimes carries the numinous."

I felt him genuinely trying to bridge our very different perspectives, but I also knew good will and love were not enough. In the container of my personal analytic work, we were grappling with a much larger issue—one I knew I could no longer ignore.

Thirteen

May

Bees and the Soul

The bees swarm in May. You know it is coming when you see the worker bees swelling from their hive, hanging from the hive porch in bee congregates as plump as ladies' purses. The swarm may not leave that day, or even the next, but the beekeeper stands forewarned: A virgin queen has been formed. If we let nature takes its course, as biodynamics suggests, the old queen will leave with her loyal workers, who have gorged themselves on honey, very soon. There may be a second swarm or even a third within a week or so, and then another queen leaves. Hopefully, not too many swarms will leave, taking your strongest bees with them.

When a swarm prepares to leave the hive, it is a magnificent sight. The bees become more and more active, then suddenly the air around the hive is alive with bees, spiraling, spiraling up, into a branch nearby. Within ten minutes, a density of bees, which Natalio calls a "baby," will be hanging from a branch. If the branch is within reach, Natalio says the bees want to stay with you. If it is too high, they probably do not.

Natalio says this is not a time to be greedy. Catch the swarm too soon and you confuse the bees; then they are more likely to leave. Better to wait a day before shaking them from their branch into a bucket, making sure to get the queen. Then gently pour them into a new hive box.

Once I took pictures of the swarm leaving the hive and collecting into the "baby." I even went to the tree during the night and took a photo with a flash. The next morning, Natalio used a feather to gently brush the "baby" into a hive box, but by early afternoon the hive was empty.

Natalio was angry and scolded me. He said the bees were doing serious work, and to stand around and watch and then to take pictures did not show respect. This was one of Natalio's many lessons: we are not impartial observers; our presence has an impact on events, our state of mind influencing the outcome.

Rudolf Steiner said that a swarm of bees looks like the human soul leaving the physical body. He said the ego holds our many parts together until death, and then we go to pieces and each one flies away from our physical body.

I had a visceral reaction when I read Steiner's words. In my mind's eye, I saw the "baby" hanging from the tree branch, perhaps eighteen inches long and a good eight inches wide. I knew in my gut that Steiner was right. The one time I saw a soul leaving a body, I can tell you, it did look a lot like a swarm leaving a hive.

It had happened on a spring morning. As my husband's family and I stood around the casket of my husband's nephew, full of incredulous grief, Donald's niece read a poem by an anonymous author.

> *Do not stand at my grave and weep.*
> *I am not there, I do not sleep.*
> *I am in a thousand winds that blow, I am the softly falling snow*
> *I am the fields of ripening grain. I am in the morning hush,*
> *I am in the graceful rush of beautiful birds in circling flight,*
> *I am the starshine of the night.*
> *Do not stand at my grave and cry, I am not there. I do not die.*

We stood under an old, thick-barked cork oak in the rambling cemetery, suddenly acutely aware of every bird song. It was then I saw "it" —a silver-grey cloud the shape and size of a honey bee "baby" lifting from the middle of the casket and moving up until I could no longer perceive it. I was awestruck.

Jung talked of the ego as being one of multiple centers of consciousness within one's psyche. We are constellations of *scintillae*, or soul sparks, he said. Our ego consciousness dominates a good deal of the time in the first half of our lives, but in the second half the work of the ego changes: to prepare for the last "swarm."

The Taoists say that until we have formed the *subtle body*, that agency by which we consciously survive death, we are dependent on our loved ones still in flesh to "re-member" us until we reincarnate. Showing respect for these "soul sparks" by means of a receptive consciousness supports the formation of *subtle body*. Goethe, Jung, and Steiner all understood this, and, in his own way, so does Natalio, who has never read their work. Like them, his teacher has been Nature, and through him I am learning Her lessons as well.

Fourteen

Colorado Group, Last Years

The two symposiums I attended after Poul's death were difficult. As a group, we either needed to enter a new stage or bring things to an end. Poul's death had made that clear, at least to some of us. We could no longer rely merely on the excitement generated when we were together.

We had accomplished our original purpose, having spent several years exploring the common ground of analytical psychology and shamanism. The edited papers were now with a publisher. Collectively in a liminal space, we needed to either say goodbye, which many of us would have felt as a tremendous loss, or move into another orbit, which we could neither define nor agree upon. This was not unlike the dilemma I was suffering in my analysis.

The first year after Poul's death, we met at a ranch in Montana. Again, the energy was high and people were falling apart. When this kind of energy is tapped, anything not related to the Self is shaken, as depicted in the Tower card in the tarot. During the symposium, I dreamed that I was to build a fence to contain dogs. *The area is rectangular, with the proportions of the golden mean. My husband Donald asks me to lay out the dimensions and then he will build the fence. I realize that it doesn't have to be as large as I am making it, that I can close off one end of the rectangular area, still in the proportion of the golden mean, and it won't be as disruptive of the landscape.*

Then I travel in an open tram car through a fun house-like tunnel into the earth. I meet sleek black dogs and am concerned that I will be attacked. I go deeper, and still they run at me from the direction in which I'm headed, but I'm alright. The dogs belong to the owner of the property.

These many years later, I recognize how much I was working to contain a certain kind of very dark, instinctual energy. Dogs are man's best

76

friend under good circumstances, but they also guard the gates to the Underworld. Black dogs are related to the diabolical and sorcery, the shadow side of healing. There is a descent in this dream, and in the tram car I am at the mercy of where the tracks take me.

I think our group was on a collective descent that was quite danger-ous, partly due to the lack of recognition of the seriousness of the situa-tion (fun house tunnel). Certainly we were not in a particularly good relationship with instinct. In fact, the medicine man who was leading the sweat lodge (not Pansy this year) actually burned several of us, and still we participated in the lodge with him. At least two of our group experienced heart pain during the sweats.

The earlier part of the dream, in which a fence, a boundary, is being built, implies a containing partnership with my animus (and husband). It does not have to be a large area, which would be unnecessarily dis-ruptive to the natural landscape, only one in proportion to the golden mean or, as Jungian analyst Ed Edinger says, "...the ego's proper rela-tionship to the Self."[27] I—and maybe we as a group—needed ego and Self in the right proportions in order to contain the instinctual and shadowy energy of the sleek black dogs. Now I see that this dream an-nounced a personal descent on a track into the Underworld and offered guidance for containing the dangerous energy to be encountered there.

The second year after Poul's death, another incident foreshadowed my own personal descent. On the first evening of the symposium, Pan-sy led us in an opening ceremony in which she smudged each of us us-ing sage and an eagle feather. One of the members carried Pansy's staff as Pansy walked around the circle, stopping by each of us to whisk the smoke about our bodies, cleansing our auras. At one point, Pansy hand-ed the staff to me so that she could smudge the member who had been holding the staff. Afterwards, we were to hold hands to close the circle. I was setting the staff down in order to hold hands when, with a shock, I saw a stuffed eagle's head on the end of it, its beady eyes staring into mine.

---

[27] Edinger 1985, p.198.

I felt as if lightning had struck me and almost fell over. The power of the staff itself, and of looking into the very large bird head on the end of the staff, pulsed through my body. I became confused, at once pulled to take care that the staff didn't fall and also wanting both hands free so I could fully join the circle. Somehow I managed to lean the staff securely against a nearby chair and then grab the hands on either side of me, but the incident stayed with me, the eagle's ruthless eyes confronting me with my need for parameters and protection when doing soul and spirit work in a group setting.

* * *

Not surprisingly, the entire symposium that year had a depressed energy about it. Since we had met in Montana the year before, this was the first time we had convened in Winter Park since Poul's death, and we were struggling.

Don gave his paper on the tier system, the stages of analytic work and corresponding transference issues, which ignited discussions about our differences. The group was split. Some analysts and therapists had active patients present at the symposium. Others felt this negatively impacted the energy of the group. There was a contingent that wanted to do group process as a part of the symposium. Some wanted to stay mainly with the experiential work of the sweats and ritual. Discussion after papers was polluted by attacks and arguing. The symposium had become unsafe.

Yet even with all the disruption, my sweat lodge experiences deepened. Before dawn Saturday morning, I had one of my strongest experiences ever. In the intimacy of the hot and dark lodge, the ancestors came. I felt their closeness as I prayed. I can still smell the sweet sage permeating the air. The hard, dry dirt soothed my legs as sweat soaked through my dress. The Golden Serpent entered me, resting her large golden head in my heart. Then pictures began to flash before me as I heard the voices of the ancestors: "There's something we have to tell you."

The message I received still breaks my heart. I listened and cried, shaken to my core. I saw my sister being taken from me, and was instructed in what to do to prepare for this. When I left the lodge to go to breakfast, I could not stop crying. The ancestors' parting words to me were to leave my heart open the next days and they would keep speaking to me.

I later learned that my sister had received a terminal diagnosis of an autoimmune disease, which later went into remission. Thankfully, she is still with me. However, I did lose two people very close to me within a short time after the symposium: Ellen, my Southwest roommate, who died of lupus, and then, within the year, unepectedly, Don Sandner.

Whenever I have experienced the presence of the Golden Serpent, I have been in Dream Pond consciousness. The snake, especially when shown swallowing its own tail, symbolizes this state of at-onement, a state in which we are open to the energies of spirits, living and dead. If we have not yet developed an ongoing relationship with the dead, the ancestors, we may not experience this kind of love, but instead be stricken with grief for lost loved ones.

In my own experience, the Dream Pond state is one of great love and open heartedness. After the sweat lodge, my heart was wide open and I felt receptive to the wisdom of the ancestors. I felt strong and clear, and it was in this expanded state that I confronted two group members a few hours later, one who had been a problematic analyst to me in my early years at the Institute, Abram. Both had recently behaved toward a third group member in a way that I perceived as unfair. In the wake of my powerful sweat lodge experience, I felt a special energy flowing through me. The Golden Serpent rested in my heart as I spoke to the two members.

Pansy often told us to walk carefully the four days following a sweat lodge ceremony, and I believe that this should be honored. Group process and confrontation are beneficial, when done correctly, as is the sweat lodge experience. Mixing the two, however, was disastrous that day.

When I confronted the two analysts, I was met with a vicious attack. I remember being called evil and asking in astonishment what was evil about my actions. I don't recall much else, except that everyone else observed the exchange in stunned silence. Finally, someone bellowed "Stop!" and we all fell quiet. I held myself together until I was back in my room, and then the crying began. Later that evening, I was again confronted by Abram, this time during a wild dance, part of a psycho-drama-type exercise. Several had donned costumes for this exercise, and Abram was dressed in a mask and a skirt. He presented me with a cat mask, suggesting that I wear it for the course of our conversation, which I refused.

Feeling both seduced and manipulated, I argued with Abram until I heard the Serpent say, *Get out of here now!* I ran from the lodge into the cold, jet-black night. Sitting on a rock, I listened to the odd music emanating from the lodge and watched heads bobbing in dance through the vast windows. I wished that I were home.

I suddenly felt a shiver, sensing that Death was in the darkness behind the boulders. Again I heard the Serpent, this time telling me to go back to my room. One of the more fragile members of our group, the one whom I had defended with Abram, told me later that she had been walking along a precipitous path in the dark at this very time, contemplating suicide.

Much later, my training analyst Mary Jo told me that it is important to reserve the right to say no to the Golden Serpent, or the Self. Just because one has access to clarity does not mean one should speak it. We must listen to and respect the timing of things and our own personal state at the time.

This is one of the greatest lessons I eventually gleaned from this underworld experience, the true significance of the Golden Mean, the relationship of the ego to the Self. The ego needs to stay on board and make wise decisions about what is to be done, without getting inflated by the energies of the Self.

If I had it to do over, I would not confront Abram, at least when and how I did. In my wide-open energetic state, I invited attack and felt shattered afterwards. I think that we all were shattered, really, but I knew it in a very personal way. I also knew that it was no longer safe for me in this group and that I would never return. We were drawing forth energies without a safe container. The experience of our collective descent left us too vulnerable to the dogs of death.

Someone at the symposium later commented, "This year you were the group sacrifice." I have contemplated that. For the group, this may be true, but for me the event was part of a larger initiation. I wrote in my journal, "...it has taken a toll. I am rearranging inside. I'm bigger than I was. But I am soft, and vulnerable. I'll be stronger later. I hear the snake, though. It comes. It gives me strength and grounds me. I have been torn down in order to be remade. The crying yesterday: do not put it on personal complexes alone."

The night I returned home, I dreamed that lightning struck a large old tree, splitting it into two parts. Although the left side was green and thriving, the right side had fallen and was dead. I knew then that a group experience I had once cherished was over.

* * *

During the years of becoming attuned to my energetic self, I found two healers, a homeopath, Nancy, and hands-on healer, Greg. Upon returning from Winter Park, I called them both because I could not bring myself to eat. They both had cancellations that very morning and could see me.

After carefully listening to all the details of my experience, Nancy said, "This is snake energy. You were strong enough to carry it (in the positive form) and the two confronted analysts were strong enough (in the negative form), but the group was overwhelmed." She said the fact that I had been clear and empowered by Spirit was positive.

After some deliberation, she decided that the golden snake in my heart chakra sounded like cobra. "The cobra is a spiritual snake and protected Buddha," she said. She gave me *Naja Tripudia*, cobra snake venom, as a medicine. She said something major had changed and that I was beginning something new.

I saw Greg shortly thereafter. "Next time you go to a conference," he quipped, "take your armor."

"I am not going again," I said.

"What do you mean?" he asked. "This is your Dharma, your Karma. You will meet it time and again. You don't hold a mirror up to someone and then pull it back to see how they are doing. Back off, let them integrate what you said. You really did okay otherwise."

Within an hour, I felt more myself. Again I could see the light in the vines, in the grass, in the goats we raise. I was weathering a great storm, one that continued for five years and would not stop until it had leveled almost all my mentor relationships.

Fifteen

June 10

June 10 is generally the beginning of the lavender harvest for the flowers we dry. The work begins before sunrise, as we try to get the day's harvest in by 10 a.m., noon at the latest. The life forces in the plants rise in the morning hours, we have been taught. Later in the day, they sink back into the earth.

We have the system down now: Natalio cuts the stalks in handfuls and his son Albion puts a rubber band around each bunch and stacks it with others in groups of ten. Donald loads them onto the trailer behind our small, four-wheeled motorcycle and drives them to our homemade dryer, a cargo container that Donald has outfitted with heater, fans, and racks. He hangs the bunches, each the size of a pound of spaghetti, upside down to dry for two days.

We dry the lavender in the dark and in heat that does not exceed 120 degrees, then take the bunches down one-by-one and clean them by rubbing their stems together, knocking out the leaves and bind weed that may have dried inside the bunch. We can handle about a thousand of the bouquets at once, and we have a window of about five or six days to get the dried crop harvested. If the flowers open too much, the bouquets will shatter.

We have planted four distinct areas on our ranch with lavender and Natalio watches the crop carefully, taking note where it is flowering out first. We cut the lavender when only a calyx or two has unfolded its purple petals to the bees. Under this regimen, we can only get about three or four lockers full, so it is important to make sure the lavender is drying efficiently and that we start as soon as the first calyxes unfold.

For the last two years we have sold out long before the next harvest, but it has not always been this way. We planted the lavender at the suggestion of our viticulturist. Some of our vines were struggling and he had just been to a seminar on lavender. He said we could sell every stem for three or four cents. With dollar signs in our eyes, we planted the crop, thinking that each plant would yield from $15 to $22 dollars. With 3,500 plants, that grosses 52 to 77 thousand dollars!

We soon learned it would not be such easy money. Our first year of harvest, we lined up a buyer who said that he would take the whole crop, although warning us it would not be at three to four cents a stem. He had asked us to bring him a bunch of the grosso, our main cultivar, in early June so he could determine the precisely right timing for harvest. I never will forget the day Donald and I took a couple of bunches to him and he dismissed us, saying, "Uh, I don't want grosso this year!" We had 3,500 lavender plants, each with 500-700 stems, and in ten days they would begin to bloom. I thought I would throw up, but Donald, used to the grape industry, informed me that harvest often presents emergencies, so not to fret. We would simply go into high gear and find other markets for the lavender.

Although we had planted the crop for profit, Donald and I were by now totally seduced by the lavender experience. Walking into the purple sea, you become aware of a buzzing presence. Bees are everywhere! Of course there are honey bees, but there are all of the native bees, too: orchard bees and yellow-faced bumble bees, carpenter bees and mason bees. The longer you stand in this hive of a lavender field, the more you forget why you entered in the first place. The earth draws you to it like a magnet and you sit, mesmerized by the den of buzzing.

And the fragrance! That first year, the scent of it mellowed us all out. At night, we opened our bedroom window to the lavender field we had planted to the east of the house, the pungency of the flowers protecting our sleep, and when the sun rose, the purple was so intense it hurt our eyes. I sometimes spread our laundered sheets over the rows of plants so our bed could be perfumed with this grace.

However seduced we were that first year, we were also facing the reality of a lot of unsold lavender. So having signed up for the Friday evening Chef's Market in Napa, I loaded our Jeep with a table and checked tablecloth, an umbrella and chair, and tubs of the freshly cut lavender. On the way downtown, I had a talk with the lavender. "Lavender," I said, "you are beautiful and seductive, very sweet and useful, but we need your help! You are going to have to help with the marketing, or (and I don't say this as a threat but only a fact) we are going to have to pull you! We cannot afford to grow a plant that requires so much labor and has no market whatsoever." Then I let it go.

The Chef's Market meets from Memorial to Labor Day. On our first evening as vendors in mid-June, the street was lined with farmers selling their produce, local onions and honey, flowers and greens. It took a little while to find the market manager to direct us to our spot in front of an old stone building. After setting up the table, my husband and I pulled out the tubs of lavender. Later people told us they could smell it a block away. And people were not the only ones who smelled it. From all over downtown Napa, bees came! In great numbers, they covered the lavender.

I've noticed that whenever bees arrive on the scene, the energy level goes up a notch or so, and this was no exception. People were not sure whether to be frightened or excited, but most were the latter. Occasionally, someone would ask me to brush a bee away so she could smell the flowers. Each evening, we sold out.

That year we also sold fresh lavender at supermarkets and hotels. My youngest sister and her husband and four children visited, and we drafted them to help us harvest and sell the lavender, making lavender wands and other products to take to market. Fortunately, we were able to sell a good portion of our lavender fresh that year.

Then, in early July, when the lavender was fully blossomed out, a reporter contacted us about doing a story on our organic ranch. When he arrived, he too was seduced by the lavender and the bees. At the end of the afternoon, he noticed that some of the bumble bees stuck their proboscises down into the pistils of the lavender calyxes to spend the night. He arrived before sunrise the next morning, and the morning after that, to photograph the bees waking up, backing out of the calyxes, and flying away. He shot the intense purple at sunrise, and the lavender being cut into burgundy sheets for distillation for essential oil (for we had decided to distill what we could not sell). He spent some fifteen hours photographing, and his beautiful images ended up in two-page articles in two Sunday papers.

This was our debut in the lavender market. We sold out our crop that year, and we gave thanks to the lavender and the bees for being our marketing buddies and very good friends.

Sixteen

Descent

I returned home from Denver after the disturbing 1996 symposium and set about finishing my final paper for certification as an analyst. I was hardly in the most settled state of mind for meeting this task. I had a lot of unfinished business with Don Sandner, my other control analyst. The incident with Abram had stirred old feelings. Early in my candidacy, I had been hurt by Abram with some of the same arrogance and misuse of power that he showed at the conference. Had he been confronted by someone in authority then, I imagined, the situation would not have progressed to this state.

Don should be the person to confront Abram, I felt, and he simply wouldn't. He admitted, as had others, that the man's behavior was extreme, but he also explained to me what may have set Abram off. Once he said, "I am not going to throw Abram out of the symposium!" I replied, "That is not what I am asking for! I want someone who has authority to confront Abram about what he is doing." Don and I had never had the kind of head-on collision we experienced in the wake of this symposium, which—though we didn't know this then—would be the last one either of us attended.

For me, this was another step in Don's assuming human proportion. Before, I had experienced his presence as creating a magic circle in which nothing could go wrong. This time, in my eyes, he failed miserably. Although I fought not to, I had projected omnipotence onto Don. And although our contact was within accepted bounds, the archetype of incest was also operative. I loved feeling like the valued daughter who got special favors from him: his taking my group on the trip to the Southwest, then my being invited to attend the symposium.

Yet even knowing what I know now, I would still attend. These potent gatherings changed my life, initiating me onto a spiritual path. Not understanding the tier system at the time, I was reluctant to acknowledge the importance of the transferential aspect of my relationship with Don. If it became too conscious, I feared, I might lose him and his guidance on my spiritual journey. If I admitted a transference at all, I assumed that it had to be contained within the same kind of boundaries as those of my personal analysis.

In my own often difficult way, I discovered that the *nature* of analytic work defines the necessary boundaries. While I cannot imagine attending the Winter Park symposium with Benjamin, my relationship with Don allowed for our work to expand beyond the consulting room. The group experience had provided fertile ground for growth, however painful this particular aspect was.

Our work together entered the third tier, in which the subject of the transference becomes more of a teacher-mentor. "The importance of this stage is that it often involves someone other than the analyst as the object of transference," Don wrote in the paper he gave at this last conference. "The terms of transference relationship are often different than those in analysis proper… Nothing is taken for granted just because the teacher says it."[28]

I prepared to present my final paper and control case in November 1996 to the Certifying Board. Despite the difficulties I was experiencing in the aftermath of the symposium, I enjoyed writing the paper and felt more than ready to complete the process.

* * *

During this time, Donald and I were building a house. When my sons and I first moved to the ranch, Donald and I knew that we had to build, something he had been planning before we met. The 120-year-old pioneer farmhouse with its two very small bedrooms, hallway-like living room, farm kitchen, and no closets, was adequate for a single

---

[28] Sandner 2006, p. 23.

man. But now, for the four of us plus two large dogs and two cats... well, it was a bit snug.

As an architect, Donald utilizes a proportioning system that was common in the times of the classical Greeks. He also does his creative thinking and work in the early morning hours. I often found him up at 2 a.m. using a string and piece of chalk to calculate the radius of the circle that would generate a gently vaulted ceiling in our master bedroom. The center of the circle which defines the curved ceiling above our bed now rests 14 feet below in the ground. As Donald and I sleep I know we are held within a sphere imbedded in the earth and arcing over us toward the stars.

Needless to say, the building project was a predominant focus for both of us. Donald decided early on that the house should be solid concrete. We both loved the look of thick adobe walls, and he no longer felt good about using so many trees in his building. He did not like the engineering of rammed earth or straw bale homes, but was comfortable with concrete, having been involved in building hospitals over the previous fifteen years. More importantly, for years he had had a series of dreams about abandoned concrete factories in the back of the property. As he integrated the meaning of the dreams, he chose concrete as the appropriate building medium for our home.

Because we live in earthquake country, Donald had had the house engineered. This involved sinking piers nine feet into the earth and tying them into a grid of rebar before pouring the foundation. Our driveway was three-quarters of a mile of dirt road going up a mountain, and the concrete trucks could not pass each other on most of it. One truck had to wait at the bottom while the one at the building site poured out its contents.

Due to building permit delays, we broke ground in August, racing against time. We really wanted to get the concrete poured before the rainy season began in earnest. Once the floors were poured, the mason would lay two wythes of concrete block, and then the cavities of the block, reinforced with rebar, would be filled with concrete. Again, we needed those giant concrete trucks.

The floor was poured in three parts: the basement, the master bedroom and kitchen wing, and then the living room and bedroom wing. The masons followed close behind each pouring to begin the wall construction on that area. Twice, masons realized that the job was beyond their abilities and quit. Each time, the project was delayed two or three weeks while we found another team. By mid-November, we had only two parts of the floor poured, and only a small portion of the masonry walls in the kitchen-master bedroom wing had been started. The rebar web was ready for the last pour, scheduled for the Monday after my Certifying Board meeting to determine the outcome of my analytic training.

I remember seeding the slopes around the house with a native fescue the Friday before that Saturday meeting. The sky was dark, the clouds laden with rain. I left for the Institute on Saturday afternoon and within a mile of home splats of rain began hitting the windshield, then picked up speed until the windshield wipers couldn't keep up. Upon arriving to the City, I sat in my parked car, the windshield running rivers of rain. I felt odd… and yet, oddly, not anxious. In retrospect I wonder if something wiser in myself knew it was best, especially on this day, to accept outer conditions with equanimity.

I waited some time in the Institute library before the head of the Certifying Board came to get me. As I walked into the room, the six analysts comprising the examining Board—four from our own Institute's Certifying Committee and two from other Institutes—stood to greet me, then we sat in a circle.

Annual meetings with the Certifying Committee had always been lively. Often I came away with something important from the experience. I expected the same this day. After the initial greeting, I was overcome by a sense of "deadness." The Board's questions, which were meant to jumpstart the discussion, went nowhere. There was no lively energy present.

What was wrong? "You seem tired," I remarked at some point, trying to find some footing. "Oh, no," someone answered. "We've had Diet

Coke!" Soon they were actually *talking* about deadness—my deadness.

To make matters worse, the questions of one of the analysts, who had just come onto the Certifying Committee of our Institute and was a close friend of Abram, evoked a paranoia in me that I could not, or would not, address directly with the Board. Was she deliberately bringing in Abram's issues?

The longer we could not find common ground, the more I panicked. I knew that I would not be certified under these circumstances. All I could feel was the deadness in the room. In the absence of any confidence in my perception, I could only interpret this psychologically as a personal fault. Maybe I *was* lacking.

After meeting with the Board for an hour, the candidate is asked to leave so the members can deliberate. As I sat in the library for the half-hour of this deliberation, I could hear the rustling sound of the Presences across the room by the cassette tape library, the Presences that I had come to know so well during the drummings and sweat lodge ceremonies. *We're sorry*, they said, *But you would have been too bonded to this place were you certified now.* The Board told me several minutes later that I had not passed.

Plunged into shock and humiliation, I left the building as fast as I could. It was dark now and still raining. I called Karlyn, a close friend and fellow candidate in my group, and told her I had failed, then drove to Karlyn and Dick's home. Donald was there. We had planned a celebratory dinner. There was stunned grief as I recounted the meeting with them all. Later that night, as I drove home alone on the slippery black highway, I remember hearing a nasty, inner voice: *You never were up to it anyway.*

The finishing of the house foundation was, of course, rained out—just as I had been. Actually, it happened twice. Each time, we used tin cans to bail out the water that stood in the troughs. Oh, the dreams I had during this period! I was going into alchemical solution, a painfully disorienting and reactive state in which the old dissolves and the new has not yet formed.

As the shock of failing wore off, my anger surfaced. I learned that the candidate being examined just before me was dying of cancer, her death believed to be imminent. The energy in the room made sense. Death *had* been present. I was becoming much more open to these other dimensions, yet I had insufficient tools for sorting out what was what.

Hindsight tells me that I was in no condition to go before such a Board at that time, that I needed a spiritual teacher to help me navigate the inner terrain and do further preparation. Still, it would take more prodding for me to fully accept that the work required of me from here on was very different from anything I had been called to do before.

Seventeen

July
Distillation, Donald, and Kant

Distillation is a "slow, philosophic, and silent occupation," wrote chemist and philosopher Primo Levi, "which keeps you busy but gives you time to think of other things…"[29].

For thirteen days each July, Donald spends almost every waking hour in a small trailer on our property distilling lavender. Early each morning, Natalio cuts the stems short so Donald has mostly flower heads, bundles them in bed sheets, and loads the bundles onto a wagon, pulling them with a four-tracks motorcycle to a canopied area in front of the trailer. Here Donald weighs the bundles and then loads the flower heads into a large converted beer-keg still with a little water in the bottom. A turkey cooker serves as the heat source. After about 45 minutes, the still's copper coils begin to condense the liquid into drops that collect in a large glass flask. The oil rises to the top, the hydrosol to the bottom. Each batch takes about two-and-a-half hours to complete.

In this way, Donald extracts the essence of lavender. The oil is thick and pale yellow; the hydrosol, clear. There is an argument among organic growers: is hydrosol a byproduct, only water, or a product in itself? I can tell you for certain that it is a product. We distill it to a pH of 4.1, on the acidic side, making it particularly good for use in skin products and giving it a long shelf life if stored in the cool dark. The hydrosol holds the water soluble components of lavender as well as dispersed molecules of the oil.

The essence of lavender is a balance of opposites, operating on our nervous system as a tonic and a relaxant. The beauty of it is that neither

---

[29] Levi, p. 57.

is forced on you. Your body decides which it needs and uses the essence accordingly.

"When you set about distilling," Levi says, "you acquire the consciousness of repeating a ritual consecrated by the centuries, almost a religious act, in which from imperfect material you obtain the essence, the *usia*, the spirit... ...purity is attained, an ambiguous and fascinating condition, which starts with chemistry and goes very far." [30]

While Donald is distilling, I take him food and wine. I find him either reading Kant or in contemplative dialogue with the eighteenth-century philosopher. I carry the food in a basket, through bay forest and woodland oaks, and then through the sea of unharvested lavender to the little trailer near the road. As I unpack the wine glasses and the supper, he wakes as if from a dream. He tells me that he has empathy with Kant's dilemma about carrying a cultural projection of being an exceedingly rational individual. "Yet this is not so!" he tells me.

"Jung and Kant are not in significant disagreement," he continues. "Both are telling me there is present a *feeling function* in the natural state of our philosophical and psychological being. Kant allows me to know that although my *thinking* is most apparent, to the dismay of some of my children and stepchildren, my *feeling* function is still present, just not, so to speak, 'up front'."

As I pour him wine, he continues with the thoughts he is distilling along with the lavender. An architect and geometer by profession, his thinking has a spatial orientation. "You and I are not just *thinking/feeling, sensation/intuition* opposites," he says. "but geometrically adjacent, with significant overlap." He says that Jung and Kant help him understand this. We are not on a right/wrong axis; we just come from very different perspectives. His eyes sweet and smiling, he adds, "As Kant put it, I should not dismiss any one as wrong because, in doing so, I might throw out that portion I love most."

I adore this man. I pour him more wine and we eat, steeped in the essence of lavender.

---

[30] Levi, p. 58.

Eighteen

Fire

When my rage came, it stirred things up in strange ways.

Always before I had contained rage and anger in my analysis, and always before, this had worked. As I learned to recognize the personal complexes involved and ascribed the affect to earlier wounding, the anger decreased and I gained insight and maturity. I learned in this way to mend narcissistic injury, to not indulge in the luxury of inflated rage, and eventually to respond appropriately. I learned how not to be consumed by rage.

When shock subsided and anger re-surged in analysis, Benjamin was once again prepared to work regressively with my complexes. His approach only made me angrier, and I simply could not submit to it. I felt humiliated and wronged, and this time I did not want my legitimate feelings reduced to complexes. Within a month, Benjamin and I were in complete gridlock.

During this period of time, two outer events occurred that helped me find the help I needed. The first happened on the morning of the winter solstice. The storms that had begun the weekend of my Certifying Board meeting had continued throughout December, although we had had enough breaks to continue construction on our house. That morning Donald and I had walked up to the building site in what had appeared a break between storms. The concrete foundation was finished, the block walls about half laid.

The house rests on a saddle of land. To the east stretches the Napa Valley, to the west, Redwood Canyon. Far off in the deepest part of the canyon, a beautiful grouping of clouds shifted rapidly as we watched, coalescing into a funnel shape. I was mesmerized, even paralyzed, as I often get in emergencies, but Donald quickly recognized it for what it was, a small tornado, and grabbed my hand. We raced for the basement as a roar, sounding like a huge vacuum cleaner, passed over us.

We watched, curiously detached, as the funnel traced our northern property line—blue pieces of tarp and other debris rose in spirals, up, higher up, as birds floated motionless nearby. We watched as the funnel lifted and moved to a small landlocked pioneer cemetery right in the middle of our property and touched down, exploding a tree into pieces. Then it lifted again and moved to the southeast corner of our land, snapping the tops off a couple of redwoods and, we later learned, picking up a jogger who clung tightly to a chain link fence to prevent being blown away (thankfully, she suffered only a broken wrist as a result). Then the tall funnel moved out across the valley, sucking up water as it moved over a reservoir. Finally, it separated into three small funnels that withdrew into the cloud cover.

Donald and I rushed down the mountain to our small farmhouse. From our mountain vantage point, we could not tell if it had been damaged. Miraculously, the only evidence of the tornado near our house was a flat-bottomed boat impaled on a grape trellis in the vineyard.

However, we soon noticed odd things about the wiring in our house. When we opened the oven door, the lights over the kitchen table dimmed. When we turned on the washing machine, the living room lights dimmed. We called an electrician, who found that the wires that entered the house just three feet over the head of our bed were all melted together. It was a wonder the building had not caught fire.

I told Don Sandner about this in my first appointment after the holidays, thinking the synchronicity interesting. Don looked at me very seriously. "Do you think that you had anything to do with that?" he asked. The question shocked me. I didn't know the answer, and I really did not understand Don's implications. How could I possibly have anything to do with the wiring in our house?

So I decided to pay another visit to Norma T. I wanted to understand these unusual events at home. Were we being haunted by poltergeists?

Norma looked at me in the most peculiar way as she invited me into her office. I entered the room, which was still unfamiliar to me, and sat in the chair across from her. She began the session as usual, putting an audiotape into the recorder and silently reading my aura.

I had barely begun to explain why I was there when she stopped me. "I don't know what they did to you," she said, referring to my experience at the Institute, "but your energy is clear. Your chakras are open and moving." Evidently my rage had blasted some blockages. "You are different," Norma continued. "Your spirit is showing me who you are." Beginning in that moment, my relationship with Norma changed.

She told me that my *shakti*, or life energy, had affected the wiring of our house, and that I was shooting out so much energy it was a wonder I wasn't setting off car alarms (which, in fact, I had been). The tape recorder in the session kept sending out screeches, which Norma explained was due to the strong *shakti* in the room. She also surmised that the tornado had been responding to my *shakti* when it circumvented our land. "Spirit knows spirit," she said. "It goes along the path of least resistance."

She cautioned me to be careful with my husband, asking if he was tired. "When someone is running so much energy," she said, "she can deplete those around her." She sympathized with the events that had precipitated my rage, again mentioning that past life issues may have been in operation. Mostly, she applauded the energetic change in me.

I saw Don later that same day. He again told me that I needed a shamanic teacher and wondered if I could train with Norma individually. I explained that Norma was not yet taking private students, and I did not want to take classes, having had my fill of the group experience at the Institute.

During this time, I dreamed I was on a cliff near the ocean with my husband and oldest son, Jesse. *Waves crash below. A large blue-green snake with a spine of red jewels emerges from a hole in the ground. Jesse pokes at it, trying to scare it back down the hole. Instead, it slithers toward me, and I awaken with a start.*

I knew the blue-green snake was related to my healing. The red jewels on its spine suggested that this serpent, though frightening, was also a treasure of enduring value. The color red is that of the first chakra, of grounding, and also of my own birthstone, the ruby, a gem formed deep in the earth under intense pressure. My son's attitude of scaring the snake back where it came from was familiar to me, as I had felt the same earlier in my life. In the dream, this didn't work. I was to be confronted by the energy that once frightened me.

The Hindus depict *kundalini* energy as a sleeping serpent coiled three-and-a-half times at the base of the spine. Spiritual awakening involves the arousal of this serpent and its journey up the spine, enlivening each chakra as it goes. This time in my life, the dream was saying, was truly an awakening, with all its attendant shock, trauma, and upheaval.

So during those weeks and months I continued to rage—that was the closest word I could find for the experience, though it was difficult for me to classify what was happening. Analyzing feelings simply didn't work. My body temperature increased when the rush came on. Sometimes I would feel energy shooting up my spine and hitting the top of my head. I used the increased energy to work on a book I was writing. The feeling felt much bigger than what I had ever experienced before. It was more of a Kali variety of rage, She-who-destroys-*and*-creates. I honestly did not know how I was going to survive it.

I could no longer bear to enter the building of the Institute, and my love of analytic work was gone. I mourned its absence, as I mourned the excitement I had once felt in my work with Mary Jo and Don. I no longer brought any case material to either of them. I wondered if I should be in the profession at all.

One day Don asked, "Can you think of this energy as a song to God?" A song to God? *Raise up* the rage? This new idea appealed strongly to me, so instead of trying to understand the rage, to attribute it to complexes, I started painting it. I waited until the familiar pressure built up and then I let the paintbrush find the correct color and the correct movement of pigment onto paper. The images that came through were not of uncontrolled destruction, but of a woman meditating within a flame. Over time, a blue kernel appeared in the flame. For a while after each painting session, I felt a relief that came at no other time. The sense of release was similar to that of an orgasm.

I took the pictures I painted to both Don and Benjamin. Don regarded them with interest, while Benjamin did not want to be distracted by them and focused instead on what was going on between us. His attitude made me angrier. In order to contain the anger in the analysis, he wanted me to stop seeing Mary Jo, Don, and, of course, Norma, and to increase my weekly sessions with him. He said that what I was doing with the others wasn't helping.

Suddenly I knew that I was done with Benjamin's traditional format. I had been coming to analysis two or three times a week for 18 years. I needed something different, and it needed to come from within myself, like the healing paintings came from within. Mary Jo suggested that my lack of libido for doing personal analytic work might be due to a separating process with Benjamin, my need to differentiate myself from him. She supported me in this, saying that it was part of some very important work.

It was time to begin operating on my own authority. I might not have been officially certified as a Jungian analyst, but within myself I was finished. I cut down on my control analysis, seeing Mary Jo and Don monthly, and I began seeing Benjamin only once a week. I continued painting and, in active imagination, consulted the inner woman who meditated in the flame, the developing aspect of my own soul. One night she came to me enveloped in flames and said the rage had burned me clean. Perhaps this is what Norma had meant by clearing the blockages in my chakras. The fire woman told me that I had survived the rages of humiliation, of abuse, and of feeling unloved.

And so I began to see fire everywhere: in the grass, in the oaks, in the density of the bay forest, and along the road to our building site. Everything glimmered with aliveness, an extra shining. I feared that I might be physically dying.

I tried to capture what I saw in photographs. I tried to paint the fire.

In a consultation with Joseph Henderson, one of our Institute's original founders, I showed him some of my paintings, particularly from dreams having to do with green snakes, and that's when he named it for me: *lumen naturae*. The light of nature. At one point, I remember saying to Mary Jo that this energy, which I had clumsily called rage, felt in my heart chakra could even be described as love.

The night before what would be my last consultation with Don, I dreamed: *Don has been consulting at my house, but it is time for him to go home. I want to give him something to eat before he goes and offer him some soup that I served my family the night before. Then I realize that a redwood tree needs to be moved. Two trees stand together, both now large and full-grown. A man is working to tip one up because it has shallow roots, but it seems to break off at the crown. I point this out to the man. The tree is leaning and I am surprised to find that I can hold it up. I wonder if we've killed this tree, or if its heartwood was rotted and it would have died anyway. The area around the tree is covered with its seeds. In time they will come up, like a witch's circle.*

I still remember Don's response when I told him this dream. I explained that in waking life I had made the soup the night before the dream, all the time aware of the flame in the food. When we eat, we eat the energy of the sun and of the earth. When I served my family this soup, I knew that it was a holy meal, as every meal is holy. Certainly I had received spiritual food from Don. Perhaps I wanted to give something back to him before he left.

I also remember a darkness in him as we talked. He was pleased about the seeds left on the ground in the dream, which would grow into mature trees given time and the right environment. When a redwood is felled, a ring of young trees does grow about the spot the mother tree grew, in a mandala-like pattern. We talked about this symbolism, too.

As I left, he did something that he never had done before. He offered me an extra appointment before our next regularly scheduled one—an appointment, as it turned out, that he would never keep. And as I left his office that day, I was shocked by the thought that entered my head: that I had completed my work with him.

A week and a half later, on Easter morning, I received a call from a close friend of Don's. My husband Donald and I were at our house site, staining the beams for our roof. My friend's message was unbuffered: "Don is dead. He just had a heart attack and died."

Suddenly the dream and the feeling that I had had in leaving him made sense. My work with Don was indeed finished. And he was going home.

I wrote in my journal: *It is as if he impregnated us with spirit those days (in the Southwest). He gave us communion that last day, sprinkling pollen on each of our tongues: pollen, the essence of life. It was a blessing that we received from him. This initiation into being flame holders, apprehenders of the light... He gave me the controlled setting for direct work with energy, and that is a state of grace...*

Don passed not only on Easter Sunday, but also during the time that the comet Hale-Bopp passed close to Earth. The next week, my family vacationed at a ski lodge in the high Sierra. One night I sat alone in a hot tub under the night sky. In the clear air, I could see Hale-Bopp spreading her bright skirt across the heavens. I opened my heart and sang aloud to the ancestors. My voice felt strong. As I sang, I was joined by voices whose beauty was not of this world. I felt completely at peace as we sang. All was as it should be.

The sound reverberated off the water. Images of the past played before me like a film: Don's ecstatic conducting of "Rhapsody in Blue," which blared from the cassette player as he drove our group at 60 miles per hour through the New Mexican landscape. The hot and stuffy atmosphere of his San Francisco office, which I had come to think of as a psychic intensity. I saw myself sitting alone with him in his van at Taos Pueblo as we waited for the others to return from browsing in the small 800-year-old adobe homes converted to tourist shops. A short rainstorm dropped large drops on the roof as we sat in comfortable silence in the austere and ancient beauty of the pueblo. I remembered a session following Poul's death in which Don had tearfully quoted a child's story in which a dark and beautiful woman of the night beaconed to the protagonist to follow her through the open window.

After Don's death, as I sat in the hot tub under the infinite starry night and sang with the Presences, I remembered these things.

A few days later, at a site overlooking a beautiful valley and the Pacific Ocean a few miles beyond, a sleek black hearse pulled up to a curb and several men heaved out a clear pine casket and bore it to the awaiting grave. A single hawk circled overhead as the graveside service went on too long. No one wanted to say goodbye. After the casket was lowered into the grave, each of us threw in a flower, along with a handful of dirt. The sun beamed down strongly for that hour and a half, scorching our skin as it had in the Southwest desert.

Mary Jo told me that you never know if Coyote is on the path of good or the path of evil, and that Coyote may not know either. There are mysteries that will remain mysteries for us all, forever. Don had his faults and his weaknesses, but he also had great spiritual power, insight, and compassion. He was very much aligned with Coyote, that thief of light who brought the sun to humans.

Pansy said that because of Don's spiritual power, he had a spirit name, *Sunk Manito*, the Lakota word for "Coyote." She said that she had seen his spirit in the four days after his death, and that news of his death had reached the medicine people through the spirit world before any phone calls could.

After his death, Pansy composed a song in his honor stating that although Coyote walks on the periphery, he is always at the center. This is the mystery of Don Sandner.

Nineteen

Blackthorn

The next three months with Benjamin were difficult. Without the balancing influence of Don Sandner, I felt lost. Unconsciously hoping to return to common ground, I reverted to a more regressed stance to hold the strong feelings I was confronting, including grief over the loss of Don. But Benjamin had his own grief and then a personal crisis to contend with. Along with everything else we were both dealing with, he wanted to help me process what had happened with certifying, so he was quick with interpretations. I used the couch, ignored many of his interpretations, and argued when I couldn't help myself.

Finally, upon leaving my last session before Benjamin's summer vacation, I walked into the bathroom of his office, leaned on the sink, and stared at myself in the mirror. My eyes looked exhausted. I felt psychologically cursed by his interpretations. "You are not coming back," I was shocked to hear some larger part of myself say to the image in the mirror. "You are not coming back here anymore."

I spent the next month writing and re-writing a letter to send to Benjamin. Meeting him in person would be like stepping into quicksand. We could easily spend another year or so arguing. In the letter I said that I wanted to seek impasse counseling on what had happened between us, that I could not continue as we had been. I sent the letter with trepidation.

During the ensuing month, I consulted not only Mary Jo, but also Betty Meador, to whom I had been referred by a Certifying Committee member as a control analyst replacement for Don Sandner, and with Joseph Henderson. This time I described to Dr. Henderson the events of the last year that concerned my certification, as well as the difficulty of terminating my 19-year-analysis.

I will never forget his comforting words. "Creative people often have to encounter more obstacles," he said. "Don't let the shadow get in the way of doing what you have to do." In response to my concerns about terminating my analysis, he said, quite simply, "Nineteen years is long enough."

It was an awful month for me. I was struggling with letting go of my analysis, and also the decision of whether to go before the Certifying Board again in the fall. My mother was diagnosed with breast cancer and my ex father-in-law, with whom I was still close, with bladder cancer. I remember writing Benjamin that I knew that the reason that I could survive this time was because of the analysis. "Maybe this *is* separation," I wrote. "I just didn't expect it to be like this."

During this month I had a dream that presented me with the unconscious decision I was facing. *I am at a banquet. I am brought a huge fish, but it seems to be breathing. Then it talks. I notice the other people are eating their fish, but I know there is no way that I can eat a talking fish. Slowly the fish becomes a coyote. It has been buttered. I push myself back from the table, and the coyote slinks off the table and cautiously stalks the room, finally disappearing out the door of what now has become the basement of the church I grew up in. A woman at the table next to me knows that I have let my "food" go. She says that she left part of her coyote on the floor.*

I was repulsed and disturbed by this dream. Was I refusing to incorporate the living contents of the unconscious in not eating the fish? Or were the living contents (the fish) at risk of being cooked by the heat of emotions so present, including Benjamin's increasing use of interpretation? In choosing not to eat the fish, I was giving the fish a chance at life, as well as giving myself the possibility of another way of relating to the fish.

In the dream the fish then metamorphosed into a coyote and escaped. The trickster was at work! Although Coyote gets himself into all kinds of trouble, in many native American myths he also creates the world. Perhaps something new was about to happen! I associated Coyote with Don Sandner, and the living contact with the psyche he had supported in our work. Soul was making a decision to relate to the living contents of psyche in a particular way, not by cooking and eating the fish, but possibly communicating with it. You can't have your coyote and eat it too! (That said, the woman in the dream, who in waking life carries my shadow qualities, evidently had eaten hers, although not completely, as some of her presumably dead coyote was on the floor. She also was pointing out my decision to me. Both ways were present: incorporation and communication.) This was in the basement of the religious structure I had grown up in, perhaps signifying that the decision was based in deep religious roots that before now had rested in the unconscious of my Christian heritage.

Benjamin called immediately upon returning from his vacation, fully expecting me to come in to discuss this decision. " I have some ideas about what happened," he said in a phone message.

I knew he had ideas, and I also knew that I didn't want to argue again. "I need space to listen to the unconscious," I told him. He countered with, "We have all the evidence we need."

When he realized that I was serious about this, he agreed to seek consultation with me, something he later said that we should have done months earlier. In late October we went through an impasse counseling process in which we met individually and then together with an impasse counselor. Benjamin apologized for insisting that there was a piece of work not done, one that he could not articulate. I again told him that I needed a feminine space for holding my feeling and experiences, one that was not shelled by interpretations. He seemed to understand this. I suspect that in his individual appointment with the counselor he discussed his own vulnerabilities activated by what had happened between us, as I had in mine. We both felt relieved. I told him of my decision to go before the Certifying Board in December, and asked for his blessing, which he gave me. I also decided to meet with him after that the Certifying Board meeting to truly end our work.

However, the meeting with the Board only confirmed that I was still too angry to deal with them at all, and again, I was not certified. (I would guess the shadow woman in the dream got the upper hand, and I cooked my fish!) "Why would you come up now, when your analysis is in such a state?" they asked. "You seem to be blaming the Board. All you talk about is how wounding your experiences with us have been." "Your paper [in which I describe a patient's work] is one of the best we have read. Your work is not the problem. We are talking about your presence in the Board meeting."

They were correct, of course. I was in a rage, a protest! So many relationships had fallen apart one way or the other, including with the Board. I longed for someone to understand my feelings, and that was not the evaluatory Board's job. My unconscious insistence was only fueling my anger. This was a hard course in identifying a deep longing to have someone understand my suffering. More than once Mary Jo confronted me with the question, "What is it like to be the only one in the room with this perception (or feeling, or thought)? They are not going to understand this."

It was the same process that I was confronting with Benjamin. We had real differences. Benjamin, for whatever reasons, could not go where I needed to go. Arguing became a kind of negative merging, and consequently, a smoke screen, to our real separateness, which felt almost unbearable.

I was very discouraged. Was becoming an analyst even my right path? Benjamin and I met during December and January. He wanted to help me, but again he was using interpretations, encouraging me to increase my weekly sessions, and suggesting I stop seeing Betty and Mary Jo in consultation. "What they are doing is not helping you," he claimed again. He said that the work with them was also siphoning off energy from the analysis, a dynamic that could have been true at earlier stages in the analysis.

I was exasperated. Betty reminded me of my intention to end analysis with Benjamin. She also underscored the importance of my work with the unconscious, which seemed to be requiring quiet alone time on the mountain. She fully supported my feeling of needing to rely on my own authority. Her statements slapped me back into sanity.

So in February, for the final time, I told Benjamin I was quitting. He just closed his eyes. We sat in silence, the sadness palpable. "It seems so sudden," he said.

"Are you upset?" I asked.

"Yes," he answered. "You took what I said about stopping seeing Mary Jo and Betty in such a strong way."

There was nothing else to do but sit with the pain of separation, sadness, and, yes, gratitude. If the symbolic meaning of the crucifixion is a submission to the force field of being pinned between the opposites, as Edinger says, I understand some of the suffering of the Christ. Slowly I was accepting the necessity of willingly taking on this suffering in my maturation and individuation process. Benjamin and I met the next week and we said goodbye to 19 years of our work together. It was the same week that Donald and I moved into our new home.

* * *

When Donald, my sons, and I moved into our new home, I crossed the threshold into the new world in several ways. The stain of sadness and grief of the many events over the last years only intensified my experience of the beauty of the mountain onto which we moved. At night, the lights of the Napa Valley twinkled to the east. From our bedroom we could watch the full moon rise over the eastern range. To the west, we gazed into the rugged landscape of the Mayacamas range. To the north was oak and bay forest, to the south, an oak savannah.

Particularly awed by the 300-year-old valley oaks south of the house, I awakened all hours of the night to their graceful branches silhouetted against starry sky. Sometimes I photographed their ghostly forms in the dark, early hours of the morning. One night just after the spring equinox, I awoke to see a spiral of stars seeming to come out of the tree directly south of the bedroom. The sky was pinkish black in the background. The beauty of this scene filled me with ecstasy.

I lived with this experience for a while, painting it, writing a poem — all attempts to put on paper the *feeling* of what was stirred in me. In a consultation with Norma T., she told me that the vision reflected a time of quickening and advised me to rest and remain receptive.

In the summer of 1998, Donald, my sons, and I traveled to England and Wales. Shortly before the trip, I dreamed that I was hunting for the home of my great grandmother Rachel. I am told in the dream that there is a tree in the yard of her old house which is a key to finding her. The leaf of the tree has five lobes and is about an inch and a half long.

The dream triggered a great sense of longing in me, a feeling that was becoming quite familiar. My great grandmother, Rachel Thomas, immigrated to America with her parents from the south of Wales in the mid 1800s, but we knew nothing about the circumstances nor the location from which they came. My attention was captured by the dream tree and the specifics of its leaf pattern. The dream suggested that this attention would help me trace my female lineage.

A few days before we flew to London, I had an appointment with my homeopath, Nancy, spurred by my smoldering anger, considerably lessened now, but still there. When I told her this dream, she suggested that I collect the leaf, seed, and root from the tree when I found it (assuming that I would!) and said she would use them to make a remedy for me.

I will never forget the day we drove into Wales: August 24, 1998. The rolling hills and mountains were a rich and vibrant green, not unlike the winter countryside in northern California. I wanted to get on my hands and knees and kiss the earth. *I have returned*, I announced to the Spirit of the Earth and to my ancestral spirits. *I am Patricia, your great granddaughter, and at last I have returned!*

We stayed in a Georgian farmhouse in the hills north of Pontegarthi. That evening, the only evening of our stay that it did not rain, the sun shown with an enchanting golden light. Our innkeeper Charlotte directed us to a 12th-century Welsh castle that had been attacked

711 years before to that very day by Edward I's forces. Walking the spiral path up to the castle, I found a small tree with leaves like those I had seen in the dream. Something about the place resonated within me, perhaps because I had been feeling so pillaged myself! The tree had small red berries. "A thorn," Charlotte said later. Hawthorn? Blackthorn? I gathered a leaf and some berries and Donald dug up a tiny piece of the root.

We searched for some record of my great grandmother's existence, but to no avail, before traveling to Glastonbury. Everywhere I went, it seemed, were "thorns." On top of Wearyall Hill stood the legendary blackthorn, a small tree that also had the small five-lobed leaf of the tree in my dream. Legend says that Joseph of Arimathaea visited these islands as a tin merchant, landing on Wearyall Hill. At that time, the lowlands around Glastonbury were flooded by the sea, so Glastonbury, or Avalon, was probably an island. When Joseph planted his staff in the ground, it grew into the blackthorn tree, native to the Middle East, not England. It is said that descendants of this tree bloom not only in the spring, but at Christmas as well. So the tree is associated with Jesus, who may have visited these parts during his youth.

At a nearby bookstore I bought *Glastonbury: Avalon of the Heart*, by Dion Fortune. The author's name had come to me through Norma T. earlier that year and, at Norma's suggestion, I had read a number of her novels. In fact, the name Dion Fortune is written on the page in my journal where I chronicled my initial experience with the star spiral and tree. I had gone to see Norma the following day, and she must have suggested then that I read Dion Fortune's books.

Six months later, this book by Dion Fortune found its way into my hand. In it was a drawing of the holy thorn tree that looked like my night photographs of the valley oaks in the meadow at home. And above the tree in the drawing was an arc of stars that resembled the spiral of stars I had seen that night.

The web of synchronicity thickened around the thorn tree. After collecting the specimens from the tree on the castle's hill for the homeopathic remedy, I put them in a plastic baggy and simply forgot them.

When I presented them to Nancy a few weeks later, I should not have been surprised by her mild disdain. "They have to be fresh," she said, "not mildewed." She couldn't use them, she said; it was too late. I took them back from her, not noticing that they dropped from my bag onto the doorstep of Nancy's office as I left. She told me later in a phone call that she had found the baggy and placed it in my file.

A year to the day after I visited Glastonbury, I sat before Nancy as she found that plastic baggy in my file, still containing the leaves, berries, and root that I had all but forgotten. I was stunned at the synchronicity of the date. I had been reflecting all week about the trip, because it had had a deep, far-reaching effect on my psyche. When Nancy suggested that we finally make the remedy, mildew or not, I agreed.

Both blackthorn and hawthorn trees are sacred to the Celts and are used medicinally. Hawthorn leaves and berries, made into a tea, strengthen the heart and clean the arteries. Blackthorn, in contrast, is used for emotional matters of the heart and for suffering. I needed both. Shortly before the previous year's trip, a blood test revealed that, although my total cholesterol count was low, my HDL was also low, a ratio that statistically put me at moderate risk for heart disease. And I had certainly suffered "matters of the heart" over the previous two years. I took the remedy Nancy made during the final months that I worked intensively with Norma. I took it, not knowing that soon Norma, too, would pass out of this earthly realm.

Twenty

Ash and the Fifth Season

In the Mediterranean-like climate of the Napa Valley, there is a period of summer dormancy that has is often called the fifth season. The grasses are dry and golden; the grapes in veraison but not yet ripe; and the lavender pruned into green mounds that will look this way until they send up stems late next spring. There is not much to do but think.

It was in August that Donald sustained what we thought was a back injury. We expected it to get better in a day or two, but instead the pain kept him awake for several nights running. He could no longer cut firewood or even lift the chainsaw. Natalio carried the recycling to the road and buckets of water to the goats on the evenings when I wasn't available to do it.

Donald's back kept getting worse, not better. His legs eventually became numb and his ankles swelled, the skin breaking. He lost his balance easily. Frightened by these strange symptoms, we visited various practitioners and received various opinions and treatment suggestions. We tried different things, but his condition would improve one day and get worse the next.

A darkening happened in my world. The ash of my parents reverberated in my body and I listened: *Life is not forever. Donald and I are aging. We, too, will die.* And I heard the question: *Who will farm this land next?*

When you farm biodynamically, this is a complicated question. You have made a promise to the land in using the biodynamic sprays, and the land and plants have become more sensitive to your thoughts and intention. To return the land to chemical farming methods is a betrayal. Donald and I were contemplating not only our own mortality, but the "death" of integrative farming practices on our beloved ranch.

* * *

Ash, according to the alchemists, is a substance from which the impurities have been burned off and which can be subjected to further processes to form the Philosopher's Stone, the Stone esoteric circles call the *subtle body* or *diamond body* and Christians have named the *glorified body*. It is this part of us that survives the death of our physical and, therefore, corruptible bodies. Jung stated that the formation of this so-called *body* is our task during the second half of our lives.

Myth tells us that Zeus punished the Titans for eating the infant Dionysus by striking them with a lightning bolt, thereby reducing them to ash. Within this ash were particles of the divine infant, that symbol of the generative life force, and from them the human race was formed.

Do I carry the divinity of my parents within my own body? Does something of their individuality linger? What intimacy has been wrought upon me! I listen inwardly again and hear my father's voice this time, saying: *Let Jesse farm.*

* * *

My father died on an August day in 2004, just before harvest. On that same day in August, but one hundred years earlier, my great-grandfather also died just before harvest. So in this month particularly, the house my great grandfather built in central Illinois and in which the family has lived continuously ever since, reverberates with disincarnate relatives. To me, the upstairs master bedroom is so loud with lights that it could be the Fourth of July.

And the smell! It is uniquely musty and otherworldly, occurring in certain spots in the master bedroom of the old family house and once in my own home after I returned from my mother's memorial. As she was dying, I asked her to let me know she was okay when she got to the Other Side, and she said she would if she could. And she did, by surrounding me with the scent of the old Illinois homestead in my California kitchen!

It used to feel to my siblings and me like the dead were loitering eerily in their old haunts, but now their presence reassures us, their successors. Life and death are mysterious. The dead are all around, and their distinctive scent reminds my family of this. We call it *the smell of the ancestors.*

Our sense of smell comes from one of the oldest parts of our brains, and we mammals are what naturalist Lyall Watson calls "supersmellers." We are also scent factories, besting even the flowers with the odors our warm bodies exude. We each have a unique olfactory signature that bloodhounds can readily identify, days and even months after we have been physically present in a particular place.[31]

We slough off forty million cells a day, some drying up, some persisting for possibly centuries. As a result, we all live in an invisible sea of such cells, the Divinity of Dionysus dancing among us and through us with every breath, collecting more of himself as he goes!

Is this the akashic record from which Norma T. claimed to "read" details of past lives, or the vast expanse of Jung's collective unconscious? And is it some ancient part of our brains, with its potent ability to perceive odors, that informs us of these "matters"?

* * *

It took some time before an EKG revealed that Donald had suffered a silent heart attack. He embarked on a new exercise and diet regime and finally controlled his blood pressure with medication. It appears we have been given more time together.

Every August, though, I can't help but contemplate life and death, the seen and unseen, and the pervasive presence of ash.

---

[31] Watson, p. 215.

Twenty-One

Bringing Down the Moon

There were two consecutive blue moons during the last year of the last millennium. I began my formal training with Norma T. a few days before the first of them. Norma identified herself as being attuned to the Blue Moon Goddess, so the timing seemed particularly appropriate.

Although I had been considering training with her for some time, I had never approached her about it. In December 1998, I was being treated by Greg for continued upset in my body and he told me that there would be opportunities opening for me in the next two weeks. When I returned home that very day, there was a letter for me from her, saying that she was accepting private students. I immediately called for an appointment.

It is hard to describe Norma's training method. She drew from the Eastern traditions and from the work of theosophists Annie Besant and Madame Blavatsky. But she was also quite familiar with Jung and Christian mysticism, and she practiced modern Wicca. In short, she drew from the old ways of relating to our earth with focused awareness and consciousness. She reminded me of a raven, picking up anything with shiny bits of consciousness in it with which to build her nest of teaching. And it was in the nest that she had built over a lifetime that I spent the next nine months.

I knew that Norma had been sick, but I did not know how sick. At our first meeting she told me that she had just received a diagnosis of terminal renal failure and had been given only a few months to live. She seemed so vital that it was hard to register this. She told me that she needed to pass her knowledge on, and that she needed to do it now.

I knew that I was receiving the final fruit that was ripening just at the end of a teacher's full life. There was an urgency to our work. We

114

met for three hours once a month with an hour follow-up two weeks later, "to see how I was holding it." In between sessions, I was expected to meditate for two hours a day and to study continually. There were some months when Norma was hospitalized and could not meet.

My analysis had provided much of the beginning "clearing work," as Norma called it, so my work with her was not psychological. The focus seemed to be on raising my "frequency," a term that confused me at the time. In fact, I had to learn a whole new set of terminology during those months.

How I understand this now is that Norma worked on subtle energy levels. Raising one's "frequency" meant increasing one's life energy, or *prana*. To her, this involved becoming conscious of our energetic selves, what other disciplines call building the subtle, or diamond, body.

Our work sessions proceeded like this: We began by chatting as Norma observed my aura. She asked how my meditations were going, what I had noticed in my body, and how my life was going in general. Once, she showed a great deal of concern about a personal situation, saying that the energy was disruptive, the frequency quite different from my husband's and my own. She said the situation would impair what she and I could do. Hindsight proved this extreme reading of the situation to be true. Although the situation did not resolve easily, when it did, I was impressed with the energetic resources once again available. I became acutely aware of the energetic impact of whom and what we chose to be involved with.

After a brief check-in, we did a ritual in which I was instructed to "drink of the cup of forgetfulness, to remember who you are." Then Norma would proceed with teachings in the form of stories that will always be with me. Now when the winds come and roar through the tops of the valley oaks in the south meadow, I know that the wisdom of the Ancient Ones is returning. I know that when I feel anxious, it is my intuition stirred, stored as it is in my nervous system. When the Lords and Ladies of Creation first formed the world, Norma explained, we had certain abilities. One of these was intuition; another, imagination. Then humans started getting a little too wild, so the Lords and Ladies stored intuition in our nervous systems and imagination in our bones. This is why, Norma said, when someone lacks imagination he or she is vulnerable to breaking a bone.

Norma taught me that everything is alive with energy, and that our human task is to develop a differentiated awareness of our own energy bodies, which helps these bodies develop. She affirmed what I intuited: that the dead are here, that we are not alone, that my own intuition is well developed and accurate. She saw that I needed someone to confirm these truths, as her rural German grandmother had once done for her, and that I needed help differentiating between actual perception and imagination, and fantasy.

This teaching and check-in would last about an hour. Next came some particular exercise followed by a period of "adjustment." The exercise varied. Sometimes it involved sitting in a black velvet tent with a darshan board on the ceiling while placing my third eye and palm chakras on a light ball. The purpose of this was to open and charge those chakras, Norma said.

In one early session, she performed a ritual washing of my feet to assist in opening the minor chakras in my feet. Three sessions involved meditations in a pyramid tent equipped with mirrors and lights. In our last session, Norma taught me to bring forth harmonic tones from a specially made set of crystal bowls.

During the adjustment period after each exercise, I lay on a massage table wired with ten speakers so that when Norma played tapes, the entire table vibrated. She placed tachyon pieces on each of my chakras and both of my eyes, then covered me with a piece of silk. As the music played, I was to rest while she worked. I would occasionally hear the sound of a tuning fork and often felt it pressed into a spot on a shoulder or on my feet. I would hear her whispering words that I did not understand. As she placed her hands on my shoulders and upper chest and I received her energy, I thought, *This is melding Norma and me together.*

In one of our early sessions, I saw an eagle. At first I saw only its head, and then I felt myself pressing my face into its feathers. I loved its scent, which smelled the way I imagine my own hair smells. Next, it morphed into a bird hat that I put on. Then I was hugging the eagle, and then we were one. I realized that our *shakti* had been combined, just as had happened with Norma.

In my mind's eye, I flew to the south meadow at our home and perched in the tree outside our kitchen. I worked at seeing but that was hard. It was too hazy. I decided to visit my husband Donald, who was sitting at his architect's desk. I stood on his desk and tried to get his attention. Next I circled my son Casey, who was walking home from school. He watched me with interest. Later, I asked them both if they had had any particular experiences with birds that day. Donald said that he had stared for the longest time at a picture of an eagle on a friend's German passport. Casey told me that he and his friends had had a very long conversation about vultures.

On Norma's table that day, I felt ecstatic. I felt such love for this eagle. It was like I was making love to it with my whole soul. We were one, then we were not, then we were one again. Some of the time I had wings and was dancing.

Suddenly the music stopped. Norma removed the tachyon pieces and the piece of silk hastily, or so it seemed to me. She told me to get up slowly, but I could not get up at all. I could not even focus my eyes, which alarmed me. Norma reassured me that this was just part of the process, that we were working to shift seeing from the physical eyes to the third eye.

I could see the outline of an eagle sitting on her desk, something that seemed very funny to me. Norma, however, was not amused. She told me that I was still 80% out of my body and sent for food. I vaguely remember her coming and going from the room as I lay on the table in a state of total bliss. My eyes still could not focus, something I discovered usually happened after these adjustments and lasted the better part of an hour, and I was shocked to find that I had trouble forming words. I needed help walking when I finally got up from the table.

Only later did Norma tell me that she had feared she might lose me that day. She added that she would follow me wherever she had to in the astral plane and bring me back. She also cautioned me to eat more after leaving her office, then to go straight home and rest all evening.

This was when I began to understand some of the danger that the Colorado Group had naively walked into. Norma "followed" me psychically until she was sure of my abilities. She would sit next to the black tent or the pyramid tent and follow me with her mind's eye. Her direct observation of my aura also allowed her to see what was going on with me. Members of my Colorado Group had never been protected in these ways.

One of Norma's most valuable teachings was that she could be present with me in this other realm. She also experienced a reality there, and this brought an objectivity to my experiences.

After our sessions, I had perhaps 45 minutes of exhilaration before exhaustion hit. This was just enough time to purchase a sandwich at the deli across the street from Norma's office, eat it, and drive about halfway across the Golden Gate Bridge, heading north. When the exhaustion came, it hit with such an impact that I wasn't sure how I was going to make the remaining hour-long drive to our ranch. When I got home, I went directly to bed.

I felt so depressed the day after a particularly exhilarating session in June that I could not rouse myself. After a second day of depression, I called Norma. "Your astral body had too much fun," she said. "You are still out of your body." As she said this, I felt something like a jolt and knew that she was right, but that I was now back. After this Norma was even more attentive to the state of my aura when I left after our sessions.

And I was more attentive, too. Norma gave me the tools to ground the experience of other states of consciousness so I would be strengthened by them, not seduced into using them as a psychic amusement park, nor injured by naiveté. And by July, I was beginning to feel significantly stronger and larger. I believed in myself. I came to see that if I had entered the room for that first Certifying Board meeting in this strengthened state, I would not have had the same experience. Instead of falling into a complex, I would have remained present and confronted the atmosphere. What supported this new strengthening was my work with Norma. I was coming to terms with my psychic self. That was the level of this remarkable woman's interventions—a level my analysis never reached.

Norma taught that we have many "bodies." The four her work addressed were the physical, the etheric, the astral, and the mental. We all are acquainted with and have similar assumptions about the physical body, but we are mostly ignorant of the three "supersensible" bodies. I find this, in some ways, astonishing. Jung and two of his closest, well respected colleagues, Marie Louise von Franz and Barbara Hannah, all address the subtle body in their writings, but analytical psychology (and psychology in general) has avoided this aspect of Jung's work. The development of the subtle body is the centerpiece of eastern and western alchemy, which is also pretty much avoided in conventional psychotherapy.

In most esoteric texts, the *etheric body* is seen as being intimately tied to the physical body. At death, it is reabsorbed into the physical body or it simply dissipates. It is about the size of the physical body and is the easiest to see of the supersensible bodies. It can be seen as a whitish-blue haze following the outlines of the body, usually extending about three inches out from the physical body, depending on one's level of vitality and health. It is also a kind of bridge between the physical body and the astral and mental bodies, the realm of the opus in alchemy.

The *astral body*, or emotional body, extends further. It has colors and forms that relate to our emotional states, or complexes. A lot of early psychological work attends to the content of the astral body without ever addressing it directly. According to Norma, this is the body that is active during visions and dreams. If our *mental body* is also present, we remember the content. If not, we know that something pleasant happened, or something terrifying, but that is all. We obviously need to

both think and feel, and Norma's aim was to bring into conscious action the work of both of these bodies. She taught that the mental and
astral bodies survive death, and that both are involved in out-of-body
experiences.

One poignant memory of my work with Norma during her last few
months is that she often fell asleep during the main exercise. I remember suddenly hearing intense snoring as I was working with the meditation that she had suggested to me. At first I was upset, feeling that she
was too sick to work yet here I was, spending all this time and money
on training with her! However, when we sat down at the end, it became
clear that while Norma's physical body was asleep, her other bodies
were not. She knew where I had been as much as she had when her
physical body had been awake, as verified by our commonality of visions.

One exercise that she gave me was that of dreamweaving. In this
work, we would agree to meet at a certain time and place. This was always the Golden Gate Bridge around 11 p.m., shortly after I had fallen
asleep. I never made it to the bridge consciously, although Norma reported a dream in which she and I met under the bridge. But I definitely felt her visitation. It was like something hugged my aura, a jolt of
energy that was not mine, that woke me completely right on time on
the appointed evenings. At the same time, I often saw the color blue-
green, intense and beautiful. Later Norma told me that those were the
colors of the Blue Moon Goddess, and that she, too, presented in those
colors when in the astral realm.

The two blue moons of the spring of 1999 were in January and
again in March. During this time, Norma instructed me in an exercise
which she called "Bringing Down the Moon," in which one stares at
the full moon without blinking, as much as that is possible, for several
hours. As the muscles of the eyes tire, causing all kinds of visual distor-
tions in which the moon appears to get closer, the muscles of the ego
also tire. Lunar consciousness is brought down.

This was a difficult exercise. It was cold the first night I practiced it. Even wrapped in down I felt my body whining. The meadow glowed before me in a white wash of moonlight, alive with sound and innuendo. Stripped of everyday conveniences and hurried pace, I was open to another experience of the world. This is the stuff of vision quests, of hermitages, and getting over the first hump of it is hard. Yet if one endures the discomfort and overcomes the resistance, doors open. Other states that are always present but too faint to be perceived in the commotion of everyday awareness are suddenly apparent. Like watching the night sky from a remote mountaintop, there is nothing to obscure one's inner sight.

Norma was able to hold this other state of consciousness at the same time as the everyday state. This is the way of alchemy, and it depends on the development of the spiritual organs, or chakras. Concentration and meditation exercises help to form these organs. Norma had me work with a meditation specifically aimed at stimulating alpha, theta, and delta brain waves, which are associated with light trance states, dreaming, and deep sleep. Her aim was for me to develop a consciousness that held all of these states, along with everyday consciousness. I spent two half-hours in the middle of each night doing the meditations she prescribed.

Doing those exercises was like retrieving a distant memory or finding something long lost. One day I was questioning Norma about the different bodies and she laughed, as she often did, and said that there are numerous bodies. As she laughed, I suddenly saw Norma no longer as her physical self, but as layers in a kind of egg-shaped rainbow extending and fading far into space. All the objects around her did the same, and I *saw* in that moment that everything is one. Our bodies become one with all other bodies as they approach infinity. This sudden insight was so intense that it brought on a wave of nausea. The paradigm of how I viewed life was changing dramatically, and I knew in that instant that things would never be the same.

My everyday perceptions began to change, too. I could now observe the etheric bodies of my patients. Two who were suffering intensely with no apparent way out of very difficult dilemmas had intensely lit etheric bodies that hovered very close to their physical bodies. Certain patients had clear color shapes about them. Again, I learned that it was not necessary to go into a trance to makes these observations; I simply had to remember that place of perception that came in bringing-down-the-moon, that way of using my eyes, and then the subtle glow around a person's body, and sometimes colors, would reveal themselves.

After the session with the eagle, I felt involved with the earth in a very immediate and sensual way. One morning I walked to the back of our property, which is the highest point of our ranch, from which one can see Mt. Diablo one hundred miles to the east and Goat Mountain immediately to the south. A deep canyon runs below, and through it flows Redwood Creek. Donald and I have placed an altar at this high point to remind us of the sacredness of the land and we always approach it with reverence. Even our first old lead goat made a point of standing upon its center stone with each visit. Once, a coyote placed his scat on the stone. Other visitors leave rocks and feathers. Pansy had me place flags in the colors of the four directions there.

This particular morning I approached the altar as usual, feeling the energy rise within me, like one feels with a lover. I felt Guidance: *Hold it. Hold this energy. It is energy that could become sexual or aggressive. Held, it becomes love. Not love of anything in particular, but love of life itself, and all that carries life.* My *shakti* was mixing with the land's and I did feel love of the land, and of our earth, in my body. There was no separation.

Now when I see the brown leaves of coastal oaks suffering from the fungus that causes sudden oak death syndrome, I hurt. When I see the deer moving with stealth through the open meadow in thick fog, I feel their provisional sense of safety. When the rains finally come after months of the normal summer dryness, it is as if I have taken a drink after a long dry spell. Daily walks on our ranch are perhaps my favorite times of day. The land and I exchange *shakti*. I become larger and know things; this is how the land speaks.

Norma explained that in this expanded state we can learn many things. Someday, I hope that we practitioners of the psyche will more consciously incorporate this way of knowing into our training and work. I think it is what drew many of us to Jung's work in the first place, and it is the unconscious underpinning of what we do, but we are not trained to consciously use it, as I was with Norma. This limits our work—and our lives—more than we imagine.

Twenty-Two

B.B.

Soon Norma's training permeated every aspect of my life. I was spending at least two hours a day on various meditations and exercises she had taught me. My perspective on the world shifted as I gained confidence in my intuitive and psychic ways of perceiving.

This led to a radically different approach to a problem that arose in the fall of 1999, when Donald and I were informed by one of our winemakers, Larry, that he would not be purchasing our syrah grapes that year because he didn't believe they would have time to ripen.

The grapes in question were growing at the top of the Mayacamas range, at around 1600 feet. This 22-acre vineyard had been on the market for some time, as it had not been producing well. Even the newest vines, planted only a couple of years before, were not thriving. Large portions needed to be replanted just as Donald was considering retirement, not a good time to invest large amounts of capital. Two households lived on the vineyard property: Christino, Donald's vineyard manager at the time, and his family; and a woman artist I will call Marley.

The vineyard included four types of vines: chardonnay, zinfandel, pinot noir, and syrah. Each variety ripens later than those of its kind at lower elevations, but this year the syrah was especially slow. It was already late September; the leaves on the vines were turning, the days growing shorter and cooler. There seemed to be no way that these grapes would reach the sugar levels of 22 to 24 brix that is needed to make good wine. This situation threatened a financial loss that we could ill afford.

As had become typical in my life, it was a healer, Will, who pointed us toward a solution. I told Will of our problem during a chiropractic appointment, and he mentioned that a friend of his who had had the same problem had hired a consultant to ripen the grapes. This is how I came to meet the man Natalio would later dub *Bruja* Bill, or just B.B.

B.B. arrived for our first meeting at the ranch in a new Toyota double cab pickup. In his early fifties, B.B. was of medium build, quick in his movements, and confident, even brash, in his claims. We paid his $3000 fee for ripening the crop, rationalizing (naïveté again) that he must be successful to drive a new truck. We were to find out that he had a trickster side, as do many shamans, but that he was also gifted. Norma told me that we were lucky to have such a high-level shaman to work on our land and eventually I came to agree.

The plan was this: B.B. was to spray a series of ten special solutions, biodynamic preparations, over the next two weeks in order to balance the energies of the plants and the vineyard. Once this was accomplished, the grapes would ripen to where they should be at that time of year. B.B. would use the following sprays, in order: premix, barrel compost, and horned manure or "500" to work with earth forces and soil building; horned quartz or"501," a mineral source of silica, for supporting solar forces; oak bark compost prep to balance the silica with calcium; horned clay (one of B.B.'s own preparations) to balance the earth and solar forces; some fluorescent green liquid to balance the barrel compost, in which he composted the six biodynamic compost preparations; equisetum or horsetail tea, a plant source of silica, to balance the mineral source; valerian at 22 drops of tincture per acre for warmth forces and ripening; and, finally, stinging nettle basalt tea to consolidate flavors.

The day the initial spraying began, I met B.B. at the mountain vineyard. He was dressed in jeans, sweater, and baseball cap, not the typical shaman's garb, but his energy seemed boundless. He was walking the rows when I arrived, tasting grapes, calling the vines his babies. "I get very personally involved with my vines," he explained. We both marveled at the girth of the chardonnay and zinfandel clusters, six inches across, eight inches in places, and we tasted berries over and over. He said he loved the syrah.

"I bet that you get along best with the syrah and the zin," he said, and I thought a moment. "Yes," I answered, although I had never thought to differentiate before in this way. "How do you know?"

"Because they are more ethereal, sensitive." He made a comment about rationality, then said, "And you are sensitive, so there's a likeness." Later he told me that the syrah vines liked my walking amongst them.

As we walked, B.B. pointed out a couple of weak areas in the vineyard, one in the pinot that he attributed to phylloxera, a root louse that has wrought havoc in the Napa Valley in recent years, and one in the chardonnay above and to the southwest of a small irrigation pond, mostly dry this time of year. He said that there was a vortex there, and that those plants should be thriving. They were not, and he suggested that I work with dowsing that energy and changing it from yin to yang.

As we were walking toward Marley's house, he stopped short. "Does she smoke?" he asked.

"A lot of pot," I replied.

"She is negatively affecting the whole area," he explained. "But that can be fixed."

The young zinfandel needed biodynamic compost, he continued, a shovelful or two per plant. And water. He looked thoughtful. "Are they getting water?" I did not know. We checked the irrigation, which seemed okay, but he said, "Is it being turned on?" As it turned out, Donald had opted to stop irrigating these young vines a few weeks before because there was a shortage of water. The vines had closed down as a result. For the first time, I began to wonder about an earlier deprivation in Donald's life that he may have unconsciously replicated in the vineyard. I was beginning to think psychoanalytically about our growing methods and our relationship to the land.

B.B. pointed out cattails in the pond, which indicate a spring. He added that cattails are good energy, all parts being edible, and that they bring verticality. "Many of the grape canes are straining upward, versus horizontal," he said, "indicating a need for better use of the sun and the need for silica." Two of the sprays would work to remedy this.

He said there were two ley lines crossing in the vortex area, one going through the pond and down and over the hill to the northeast, and the other going toward Sonoma County. A soil problem in the old zinfandel was causing those vines to close down too soon. "Those grapes should not be yellowing for another month."

The pinot noir had blackberries and poison oak growing amongst the vines. "The message, like that of the star thistle in the chardonnay, is *Stay Off!* We need to tell the vineyard that we will take care of it, that it doesn't need these plants. The star thistle indicates a silica imbalance in the upper part of the vineyard." The pinot needed soil building, B.B. said. "Once there is a little healthy soil, it builds up quickly," he said, something that would become evident in the years to come.

After reading the vineyard like a book, B.B. pulled his cap off and wiped his brow. "This is a traumatized vineyard, by emotions, not chemicals." He looked at me, waiting for an explanation.

"Well, there *has* been a lot of disappointment focused here," I said. "Our former viticulturist, Sam, worried that the newer vines were not growing and that crop production had decreased; and Donald had high hopes for some income that would allow him to retire. Also, two sales have fallen through." I told B.B. that this vineyard was bought from the an inheritance Donald received from his family's farm and that Donald wanted it preserved for his children, intact. At the same time, we couldn't afford the expense that it was requiring. As I had many times before, I wondered how much Donald's and my own psyches were intricately woven into the health of this ranch, and if healing the ranch meant healing the ranchers.

The first day of spraying, B.B. used the premix. The premix works closely with the soil, containing as it does a large component of cow manure, along with all the compost preparations, egg shells, and basalt. In fact, the first three sprays are all manure-based and have to do with earth forces and soil building. I was surprised to see how little of the prep was used, only about 1/4 of a cup in 5 gallons of water per acre. After being stirred, the resulting solution is sprayed on the ground throughout the vineyard in the later afternoon hours.

The next day, Donald and I went up to the vineyard at noon. The pinot, which had ripened on time, was being harvested that day; the winemaker had called the night before informing us of his plans. When Donald talked to B.B. of his concern that the premix spray might interfere, B.B. commented that it would be better to wait, but it was okay to go ahead. The grapes with only the premix would have "a terra taste," he said wryly, "earthy, rather raw." Donald and I promised each other that we would never tell the winemaker we had just sprayed the grapes with homeopathic doses of cow compost!

The vines that had been treated with the premix the afternoon before looked better than they ever had, their foliage full and green. In fact, the pinot winemaker was concerned that we had irrigated, as this can increase the water weight and dilute the flavor of the grapes. We reassured him that we had not. While waiting for B.B., Donald and I examined two 60-gallon oak barrels of murky brown water and a grape bin full of florescent green liquid with a board stretched across its top. Six strings hung from the board into the water. Later B.B. showed me that the strings were attached to small packets of biodynamic compost preparations.

We left for a few hours, and when we came back at 5 p.m. the harvest was still on and B.B. was spraying the barrel compost with the four-tracks motorcycle. The pickers finished just at sunset, around 7 pm. We were all very happy. Donald and I walked all over the vineyard tasting grapes and the flavors were incredible.

That night I dreamed that we were making a biodynamic sandwich that we would feed to the vines through irrigation. The mixture included onions and snake meat. Used to honoring my dreams, and knowing B.B. was receptive, I told him of it. "I like hearing these dreams," he said, "because it helps me make the remedies." He said that the onion root is where cosmic energy is stored, so the balance of earthy snake meat and stored cosmic energy were feeding our vines.

Later I dreamed there was a hurricane as the result of what B.B. was doing. I saw stuff flying through the air. On Saturday afternoon, as B.B. was finishing spraying the barrel compost on the ground, I told him the dream.

"Not a tornado?" he asked, seeking clarification.

"No," I said.

"Well, a hurricane has a vortex, and you have a vortex here." He asked what the dream meant to me.

"A hurricane forms over the ocean, not the earth, as a tornado does." I reached into the associative state I had come to know so well in my work with Benjamin. "The turbulence is from the unconscious." I told B.B. that I suspected it was cautioning us to respect the power of what we were doing.

When Larry dropped in to check on the grapes later that afternoon, he was shocked to see that the syrah leaves were greener than they had been the week before and that the berries, once raisining, were now plumper. The brix had come up two numbers. He too wondered if we had irrigated, then realized this actually would have decreased the brix. Now he fully expected the syrah to ripen and, of course, he wanted the grapes.

B.B. was ecstatic with this news, documenting it with everyone. As we were to find out, his maverick personality made such testimonials valuable.

On Monday, I went up to the vineyard again. When I arrived at 1:45 p.m., B.B. was stirring. He had set a large wooden tripod over a barrel and was stirring the horned manure for the "full German hour." Facing east, he turned up the cassette player in his truck and the Moody Blues boomed out into the vineyard. "Very important," he said. "They also like Bach." He pushed the stirring stick toward me. "Stir, at least some. It is important for the owner's energy to go into the grapes."

As I stirred, he told me that in meditation that morning he had seen how all the nature spirits were moving down the hill and into the vineyard. "See that area over there?" He pointed to the chardonnay. "It's like the Mormon Tabernacle Choir. The nature spirits want to help, and they are so happy that this is being done."

"Do you see nature spirits?" he asked. I said that I was just beginning to, but not consistently. He pointed out the vortex, saying that this morning it was tall and narrow, rather than squat as it had been before. He didn't know the reason for this.

Before stirring the water that had the horned manure in it, he had me put my hand in it, as well as in the barrel that we were stirring. The unstirred horned manure water was cool and thick, but the stirred solution felt alive. The image that came to me was of little fish-like things, masses of them, swimming in swirls.

Suddenly, a shaft of bright light in the pinot behind B.B. startled me. "B.B., do you see that?" I exclaimed.

"You mean that deva?" He did not even turn around. "I wondered where it was. Usually it hangs out down there," and he pointed toward the chardonnay.

I can't accurately describe the potent feeling in the air that afternoon. The wind was changing, and with it the weather. The deva was gone almost as soon as I saw it, or perhaps I should say that my vision slipped. I have seen two other devas since, here on the ranch where we live. "They have to pump themselves up so you can see them," B.B. explained later. I felt as if a veil were being pulled back to reveal a level of reality I had never perceived before on the ranch.

The next morning was clear, ideal for the spraying of the horned quartz. B.B. started stirring before sunrise and completed his spraying by mid morning. Later that day, Larry took the brix reading on the syrah. It had gone from 19 the week before to 24.3 and he wanted to harvest immediately. We convinced him to wait so the flavors would be more balanced. He said he had bought a book on biodynamics and was already reading it.

The next Friday afternoon, B.B. sprayed the oak bark preparation, a calcium spray he said would calm things down after the release of silica energy from the horned quartz spray. "Marley's behavior is a barometer on this," he claimed. "If she is calmer, it is balanced. Now, in the astral, you see pink all though the field."

B.B. taught us throughout the process. "Plants yearn to be a part of a higher evolution of being, so they love to be harvested and consumed."

Again, I felt the sacredness of the act of eating. We are taking living spirit into our bodies. It also means that we are working with spirit when we grow plants.

The next day, B.B. was using the green florescent liquid that had been composting in the grape bin for several days as a foliar, or leaf spray. Christino and a fellow worker, Bernardo, carried backpack sprayers, with B.B. on the four-tracks. Christino was literally glowing. Clear white light surrounded his head. He looked more alive and happy than I had ever seen him.

And so the spraying continued, all of us impacted by it in some way. The last evening of spraying, Donald and I walked the vineyard while Christino and Bernardo sprayed. Night was coming on. Christino's young daughter walked ahead of the two men, carrying a streamer on a stick. Such a procession! .

We felt so blessed. The brix on all the grapes had increased several weeks' worth during the two-week spraying period. The pinot was in and the syrah was ready to harvest. Within a week, the zin and the chardonnay would also be ready. More importantly, our spirits were lifted. Healing was taking place, of the land and of the individuals who knew the land.

When Donald had gotten discouraged with the vines and the ranch, he gave up, putting the vineyard on the market. When it did not sell, he and I grew even more discouraged. Now, with renewed hope, we began to examine our expectations and disappointments. In farming biodynamically, one's intentions become more conscious and every interaction with the land becomes a teaching.

Twenty-Three

Compost and the Veil

After harvest we spread compost in the vineyard and along the rows of lavender, and then make the compost pile for the coming year. We also spray the vines with 501, horned quartz, and the lavender with barrel compost, a mixture of horned manure and compost preps. This is our final act of bringing balance for the season.

The work of composting deals with the realm of the veil, an interface between life and death, nutrition and decay. When the veil thins and the dead are nearby, five of the six preps are buried, that they may gain the strong earth and cosmic forces necessary to enliven decaying matter. All but stinging nettle are processed in animal organs, which intensify the effect: dandelion in a cow's mesentery; chamomile flowers in a cow's intestines; yarrow in a stag's bladder that has been hanging in the sun all summer long; oak bark scraped and stuffed into the skull of a sheep or cow. Stinging nettle's life forces are so strong that its leaves are buried directly into the earth. The sixth preparation, valerian, is a tincture made from pressed flowers.

Once made, these six preps, in quantities of about a teaspoon each per compost pile, or 22 drops of tincture in the case of valerian, are inserted into a compost pile to help the soil incorporate cosmic forces as well as making minerals more available to the plants.

My son Jesse opened my eyes by calling farmers quintessential, their observing presence critical to returning balance to our earth. As the alchemist's experiments worked on the alchemist himself, creating the gold of the philosopher's stone, farming—when done consciously—creates consciousness. The alchemist Paracelsus wrote about the importance of developing the inner eye to perceive the *lumen naturae*, the light of nature, pieces of the world soul scattered in matter. Jung wrote that the light of nature is the *quinta essentia*, "extracted by God himself from the four elements, and dwelling 'in our hearts.' The light of nature is an intuitive apprehension of the facts, a kind of illumination."[32]

I am reminded of my awakening to the other world as a three-year-old searching for the dream pond.

Thus is the veil pulled aside, revealing Sophia, the feminine face of God. Our culture's recent farming practices have denied Her living presence, have assaulted and raped Her with chemicals, ignorance, and greed. Yet it is She, Sophia, who gives form to all we know, and her essence is in all life. Oh, that we might recognize Her, developing our sensitivities to the presence of the Divine in matter.

Nowhere is this more important than in our work with soil. On our ranch we build the compost pile, layer upon layer, out of cast-off matter that yet holds the Soul of the World. We worship Sophia by collecting elements of Her from the beings who live here: our goats' manure and that of our llama and our friends' cherished horses. We layer this *prima materia* with straw, grape pumice, and lavender prunings.

Then we bury the preps in six holes evenly spaced, dug twenty inches into the heart of the windrows. Over the top, we sprinkle half of the stirred valerian solution, beckoning heat.

In this way we invite in God Herself and give thanks to our Mother, the living body and soul of Earth.

---

[32] Jung CW 13, ¶ 148.

Twenty-Four

The Seed: Passing On

My last session with Norma T. was on November 9, 1999 (11/9/1999), five days before her death. As it turned out, it was her last session with anyone. Norma arrived twenty minutes late due to a traffic backup from an accident on the freeway. I was reading a book in the waiting area when she stumbled in, looking gray and exhausted. I waited for another few minutes as she collected herself, found my file, and got some tea for the beginning ritual.

When she ushered me in, I closed the door behind us and dropped my bags in the area beside the chair. I helped her put in the tape, something she had done herself every time before but which she had trouble maneuvering this day, and she did the beginning prayer.

What did we talk about? I do not remember. She struggled with words and with short-term memory loss. I remember my alarm at her diminished state and my disappointment in the weak *shakti* flowing between us. We repeated a pyramid meditation we had done twice before. She commented that she had done it only two times before with a student, forgetting that I was that student!

Stunned and perhaps protecting her, I chose not to bring this lapse to her attention. I recall the uncertainty of her answers when I asked direct questions about some of my meditation experiences, though moments later she addressed something very similar in a more coherent way. I could not help but notice how much her cognitive function had deteriorated since I had last seen her.

In retrospect, I understand that in that session she was saying good-bye. She told me that there is much memory in our etheric, astral, and mental bodies; not so much in our physical bodies. She said, "You will just happen across something—it may be an artichoke in a grocery store—and something will come back. It all connects." She said that I had been a *kuna*, a dream or spiritual teacher, in another lifetime, which had made me unafraid of the spiritual realm. Then she made a leap to my involvement with the Institute. "Be unafraid there, too."

At the end of the session she told me that she had closed down my chakras, rather than opening them as she usually did. I felt more in my normal state of consciousness as I left her, and disappointed that this was so. As she hugged me goodbye she told me that she could not have been more proud of me had I been her own daughter.

Over the next days, I thought about suggesting to Norma that she take a break in December, wondering if she was too ill to realize she needed one. I also questioned whether or not I had gotten all I could out of her teaching—so similar to how I felt when I left Don's office for the last time.

Death has a scent. It comes and we "smell" it, intuitively, yet hope and even pretend that it means something else. The scent lingers after the physical body's death, and we use our intuition to sniff out where our missing loved one is. Like a bloodhound, I still go back to where the scent was strongest, the last time I saw Norma, remembering what I was doing the moment she passed or the moment before, what dreams I had that night, the night after.

During the last session with Norma, when I lay on the table at the end as she was closing my chakras, I saw her and Don Sandner dancing in the meadow. I wrote in my journal that they were attending "the consecration of my studio and our land." They were in ecstasy, and the three of us went from the south meadow into my studio, our hands raised to the heavens in the sign to the sky god *Shu*.

That next weekend, it was time to sow the companion plant ground cover in the vineyard. Rain was predicted and we were as usual under pressure to get it out. Due to the delayed harvest, it was already late in the season.

After Donald and I contracted with B.B. to train us for a full year in biodynamic farming, our *brujo* had developed two mixes of ground cover seed, one for the red grapes, the other for the white. My son Casey and I now mixed the seed in a wheelbarrow, about twenty different varieties for each kind of grape: vetch and Shasta daisy, African daisy and white Dutch, California poppy (only 1/4 pound per five acres!) and mustard, bird's foot trefoil and medic clover. The experience was joyful. The seeds were smooth and vibrant as we scooped our hands through them: small round pellets, flattened flea-like pellets, and larger black spheres. A community of companion plants convened in a wheelbarrow. I always enjoy working with seeds, but on this day it was particularly blissful.

We planted on a Sunday evening. Early Monday morning Norma died. I learned of her death on Tuesday when James called. "There is no easy way to say this," he murmured. "Norma passed yesterday morning."

I suppose I should not have been surprised, but I was. Death is always surprising. James told me that Norma could teach me more now from the other side, but I wanted the flesh-and-blood version.

Bereft, I went to see Greg a couple of days later. My body felt different and I worried about the effect of Norma's passing. I wondered if I was stricken with simple grief. Greg assured me that what I was feeling was not grief, that I felt different because I *was* different. He said that before she closed my chakras, she put a lot in.

"It is like seeds," he continued, "and she wrapped you up after placing all the seeds there." He said that things would just come to me, and that the nights would be particularly important, that Norma would teach me in dreams. The way to her teachings is not from the outside, he said, but to think of how she "wrapped" me and go within. "She is not gone," Greg said emphatically, gesturing toward my chakras.

So, dear Norma, you planted your seed in me. And on the morning you died, I slept in because I was tired from my planting work on the evening before. I did not know that while I was riding the four-wheeler thirty miles an hour through the vineyard, spreading tiny wildflower seeds, companion planting—that you were retching; that your blood pressure was making your ears hum; that your body, bruised from catheters, was begging for rest. It was as if God had to bring you to your knees to make you go.

Casey and I had stirred the wildflower seeds together in the wheel-barrow with our hands first, then with our entire arms, and I felt a force from deep down in the earth. I would have rolled naked in those seeds if I could. We scooped them into buckets and then into the red funnel of the planter, mixed with sand to help them scatter evenly. As we seed-ed, the clouds thickened until drops of rain threatened to make the tilled soil too pudding-like to drive on.

I had never driven the four-wheeler before, but I saw the heavy dark clouds pressing lower. Donald was driving the tractor and roller, press-ing the seeds down, and there wasn't much time. So I started the four-wheeler and drove it fast through the rows, singing the Lord's Prayer like the Egyptian pilot says his prayer just before the Great Descent. I sang loud and long, noticing that I was passing Donald up, but I didn't look to see his expression. He asked later, "Did you have fun?"

It was not fun, it was ecstasy. I courted death on the four-wheeler. I feared being impaled by a trellis, but I drove fast to stay ahead of the storm, of the pudding earth, of winter. By the time we stopped it was dark and pouring rain. The weather always catches up, in the end.

I did not know that it was your last evening, Norma, as I did not know what Greg would say. "Before she zipped you up, she planted seed in you, everything that she knew."

Wildflowers, companions to the vines: a chorus of spirits!

* * *

Over the next four weeks, I dreamed active, lucid dreams, often of Norma. One night she was showing me a mandala, paying particular attention to the center area which veered off to the left. There were three rectangles to the right.

One night a voice was dictating.

One night I was being taught about companion planting, the print on the page dating from the first of the last century.

A week after Norma's passing, I woke from a dream in which *a friend is making a joke about calling up a tornado. I caution her, relating that our land is in a tornado path, and start to tell the story of Donald and me and the tornado. Suddenly the sky is filled with missile-like objects, like a meteor shower. Then I see a voluptuous nude woman who looks like Norma flying in the heavens with outstretched wings. A man may be flying behind her. The woman looks a little like a cherubim.*

I wrote the dream down and then, in a continued state of reverie, took a walk. As I mused on the image of the woman, my eye was caught by a black sphere in the woods, a black sphere covered with what appeared to be gold stars. It rested about two feet from the ground and was caught in heavy brush. I walked closer, making my way through leafless winter poison oak. It *was* a black sphere, but there was also a white one tied with it, and they were both covered with golden stars. They were balloons, apparently escaped from a celebration. Partially deflated, they had landed here in the heart of our ranch. This happened a week to the day after Norma's death.

I felt Norma strongly in this dream and in the balloon experience. It was she who had interpreted the outer tornado experience for me in a way that acknowledged the reality of the energetic world.

A few days after the tornado dream I came into our house after my daily walk. A storm with high winds had come up while I was out, winds that heralded the returning of Ancient Wisdom. I noticed when I walked into the house that a strong wind blew inside, too, and followed it to its source. The gusts had rattled open one of our bedroom windows and then ripped it off its hinges. The inner and outer worlds mingled together, and I wondered what dream I was awake in.

This sense of comingling worlds continued. I dreamed of a bride and groom, like the black and white balloons and Jung's sacred *hierosgamos*. As I reflect now, I think this symbolized my growing consciousness of the marriage of heaven and earth, of spirit and matter, even as a tornado is a violent marriage of hot and cold. Norma always said that we are here to bring heaven on earth, to bring spirit into our bodies through consciousness, that this is how it is meant to be. During those weeks, I dreamed that I looked into the skies and saw unfamiliar stars and constellations. I realized then that I was in a new world.

And Norma was not quiet during my continued awakening to the supersensible world. Once I awakened with a start after hearing footsteps receding from the bedroom into the hallway. I had the distinct impression that someone had deliberately roused me by rearranging the covers of the bed. I have no doubt that Norma was that wakening force.

I also felt her continued protection. One night I dreamed that I had two descending snakes placed at the base of my skull along my spine. One was spiraling downward, like the serpent on the Yuroba healing staff my family had given me some years before. Again, I startled awake, thinking of Norma. The snake was moving down into the earth. I think Norma was still doing adjustments, making sure my psychic energy stayed grounded here on earth rather than dissipating through my crown chakra.

There was humor, too. One night I dreamed of tachyon's value in feeding the masses. From the other side, Norma was still promoting the use of this product, which she claimed enhances the energy level of whatever chakra it is placed on. I did finally purchase the tachyon pieces she had always encouraged me to get, in spite of the expense!

During this time, when dream and waking life intertwined like serpents on a caduceus, I believe Norma was actively showing me that she was still present. Because death by definition means one no longer inhabits a physical body, she had to rely on other means to reach me, those aimed at the sensitivities of our other bodies. Early on, her etheric body was still present, and I think many of the physical happenings in the house and on our property had to do with this. Later, as her dying proceeded, this stopped. I still dreamed of her occasionally, but the dreams were not so lucid and were not accompanied by outer phenomena. I now feel her as I have always felt the ancestors. She is there for the calling, but she is not so immediate.

Jung addressed a similar situation in a letter to a man who had experienced the ghostly presence of his brother, who had died in an accident in Africa. He wrote:

"Naturally we can form no conception of a relatively timeless and spaceless existence, but, psychologically, and empirically, it results in manifestations of the continual presence of the dead and their influence on our dream life. I therefore follow up such experiences with the greatest attention… The continual presence is also only relative, since after a few weeks or months the connection becomes indirect or breaks off altogether, although spontaneous reencounters also appear to be possible later."[33] He continues, "There are experiences which show that the dead entangle themselves, so to speak, in the physiology (sympathetic nervous system) of the living. This would probably result in states of possession."[34]

I think my dream of the descending serpents addressed this in my own body: my *kundalini* was to stay involved with the earth.

Norma, too, had suggested never letting an entity enter one's body, as some mediums do. "They might not want to leave," she cautioned. Her emphasis was on the development of the supersensible bodies and using the perceptions afforded by this work to gather information, just as one uses our usual five senses.

Once my mother told me that she was sorry she had not asked her own mother more questions about the family history, and that she had not paid more attention when my grandmother told stories. I understand now what she meant. Maybe impending death has that influence. We are afraid that if we acknowledge its proximity with such questions, we will bring it even closer.

How many times do I run through the questions I wish I had asked Norma? Those questions change with time, like a kaleidoscope. *What is your experience at the séance? What do you see when out of your body, body shapes or balls of light? After death, how long can you work in the etheric*

---

[33] Jung 1973, p. 257.

[34] Jung 1973, p. 258.

realm? In the astral? Where do you go then? And how much of your Norma integrity remains? I would ask her: *How do you close chakras? What were you really doing during those adjustments? How am I best to guard and develop what you "planted" in me?*

As I write this, I hear her laugh. "Just relax," she says. "Remember the artichoke."

I still sometimes long to speak with Norma, to be in her presence, to lie on her table and receive her healing ministrations. But she taught me one thing, above all. In the end, we can't get salvation from anyone else. We must come to it on our own.

Twenty-Five

Endings and Beginnings

My sons thought Donald and I were nuts to hire B.B. for a year of training at the flat rate of $25,000. Jesse and Casey had always worked on our ranch—helping to build our house, cutting firewood, and weeding the grapes and lavender. Even though they were on the payroll, you would have thought them indentured servants to hear them complain. They complained about the boredom of field work, about having to get up early and start work at 8 a.m., about our taking out withholding from their checks. They were certain any other employer would not treat them so heartlessly.

Jesse was just entering college, majoring in botany (and soon to participate in a composting operation at the school dorms); Casey was a junior in high school. In later years, as young adults, they both came to value manual labor, as I do, and the growing of healthy food. Jesse has his own Community Supported Agriculture (CSA) Farm; Casey worked on a film project about the history of farming while supporting himself in landscaping. But at that time they were true teenagers, wanting to keep their own schedules and to function independently of their parents.

Also typical of teenagers, they both had good noses for fakery, and B.B. had just a little too much charlatan in him for things to go well with my sons. Donald and I were still in the golden haze of the harvest, impressed with what B.B. had accomplished. The vineyard on the ranch where we lived was also struggling, and our viticulturist had quit. We needed a new strategy and B.B. offered it, so we overrode our doubts and continued down the path with him.

B.B.'s plan was comprehensive. First and foremost on his list of priorities was soil building. This involved digging a cubic yard of compost into every 25 vines over the first three years and every three years thereafter, about a cubic foot per plant per year. Cover crops of the wild flower legume mix would break up compacted soil. The boundaries of the vineyards were to be defined with plantings of shrubs and perennials, protecting against destructive insects and influences while also flavor-crafting the grapes nearby. We continued a bi-annual, seven-step biodynamic spraying program on the grapes and now also on the lavender, which we had planted that fall. (We have modified this plan, but it is still the basic structure of our farming.)

Finding the manure for making the compost was no simple matter, and it wasn't inexpensive, either. B.B. had a special formula for the grapes: we were to mix 60% cow manure, 20% horse manure, 10% chicken manure, and 10% grape pumice. This was then to be layered 50/50 with straw in long windrows, treated with the six biodynamic preparations, covered with a foot of straw, and left for a year. B.B. supervised the layering of the piles, which Casey helped with, scowling the whole time. "What an attitude!" B.B. chided.

Donald drove about trying to scare up horse, cow, and chicken manure, as well as straw, from our neighbors. When he was unsuccessful, we paid $720 to have 40 yards of cow manure delivered, only a third of what we needed. My father, still living at the time, regarded this with disbelief. Since then we have found a much cheaper and closer source, in keeping with biodynamic protocols sensitive to carbon footprints, but it has taken time.

Early in the spring, B.B. met us at the local nursery to pick out border plants as well as some bleeding hearts for a forest garden, as B.B. called it. He picked out a clematis, two kinds of honeysuckle, and trumpet vine for the south border of the vineyard—choices a native plant consultant grimaced over later—and a magnolia for the lavender, which died within a year. He also picked out rosemary, sanicula, fleabane, wormwood, and hyssop for the north border. For the east border along the road, we ordered 20 cryptomeria plants to process pollution and a Luther Burbank cherry clone for flavor crafting, all of which also subsequently died.

He insisted we plant two redwood trees at the road, because a redwood clump on the neighboring property was lonely. These have thrived, as well as a maple tree he insisted we plant in the yard of the small farmhouse that Natalio came to live in within the year. The maple, with its valence of silica and cosmic forces, brought balance to the three-hundred-year-old valley oak (calcium, earth) that also grew in the yard. Across our unfenced road boundary, we were to build a wooden fence with cross boards, forming X's, to protect against negative energy. He also had us install a St. Francis statue between the west border of the vineyard and the forest.

Part of B.B.'s training method was to read aloud to us from Rudolf Steiner's *Agriculture*, a collection of talks Steiner gave in 1924 when he was founding biodynamic farming. I am glad that we had B.B. to interpret, as Steiner's approach was so esoteric and unusual that we needed a guide to get through each paragraph.

B.B. continued his in-the-field training. One day he informed me that our upper vineyard was a feeder vineyard for the Sonoma and Napa Valleys, so the health of hundreds of thousands of people were dependent upon it. He said that if I stood in the zinfandel and looked toward the forest, closing my eyes for 30 seconds, then reopening them, out of the corner of my eye I could see the spirits, and that there were hundreds of thousands of them. He likened it to the "Hallelujah Chorus" and said that Handel had first heard this great music being sung by the spirits. If I sat with my back straight, B.B. claimed, I could hear the chorus, too. "You hear it first behind your right ear," he said, "and then it grows in intensity up and down." He said that we had a mere quartet down where we lived, compared to what was in that upper vineyard, and that we should feel honored to be its guardians.

Once, when I was pregnant with Casey and driving through a redwood forest, I heard the angels. I knew they were singing because of Casey's imminent birth. I never heard the spirits sing in the upper vineyard, but perhaps that is because we sold it before I developed the ability. We had taken the vineyard off the market after B.B. renewed our hope by ripening the grapes, but when a realtor approached us with a serious buyer, we were glad to be relieved of the financial burden. Simultaneously, a commercial investment presented itself which would provide good retirement income for Donald. So now our farming would be limited to the eight tillable acres of our wooded ranch closer to the valley floor.

B.B. was appalled that we would sell, but not as much as Larry. Now that the syrah had ripened and been made into a great wine, Larry decided that he wanted the grapes for the full four years in the contract that we had signed with him. The new owner did not want to extend the grape contract beyond two years. Perhaps more to the point, Larry had wanted to buy the vineyard himself, but for less than what the vineyard was worth.

The sale was much more burdensome than we could have imagined. The realtor had not looked out for our interests, and in our eagerness to sell the ranch, we hadn't either. Larry insisted that we pay him for the potential loss in wine sales, about a quarter of the value of the ranch. B.B. referred us to an attorney, who concurred that Larry had written a very good contract for himself. We were stuck. We would have to make a deal or risk high court costs and possibly still lose the case. We were told that there were no clear precedents for the sale of vineyards and existing grape contracts.

We had been naïve, B.B. said. "In growing grapes you have to heal your thinking, because the vines do just what you want them to. The problem with the Syrah not ripening was this: the grapes wanted to avoid the issue of harvest and the complications in selling the land. I interceded and ripened them. Now you have to hit this issue in a more conscious way. What is your intention here?"

In Biodynamics, and in psychological work, establishing boundaries, whether physical or symbolic, is essential to the process. In establishing the border of the vineyard, we keep out what needs to be kept out and keep in what needs to be kept in. We take responsibility for the portion that we can.

Elizabeth Lloyd Mayer, a psychoanalyst and researcher into "extraordinary ways of knowing," asserts that how we assign boundaries will critically determine what we end up *seeing*.[35] In other words, establishing a boundary is a prerequisite for consciousness and defines our experience of reality.

B.B. told us that there are eight simultaneous levels of reality. In the other dimensions (or frequencies, as I told Donald, who objects to the word "dimensions"), everything is happening simultaneously. So when we perform an act, like placing a St. Francis figure on the forest border of the vineyard, we are affecting the energies on all levels. The meaning of the act may not be fully apparent to our worldly senses, but its importance may be vast on all these other levels. This is why ritual, which our culture has relegated to superstition or religious fanaticism, can work wonders.

* * *

The spring of 2000 was a time of reckoning on several levels of my personal reality. I had decided to go before the Certifying Board one final time. As the time approached, the anger associated with my first experience resurfaced. I was also very angry with Larry for causing us so much legal and financial trouble. I dearly loved our land, and his involvement tainted that.

When I thought of the land itself, I felt large. When I felt anger, I shrank. I noticed how similar this was to my feelings at the Institute. In my anger with those difficult experiences through the years, I shrank. Norma T. helped me heal this, even suggesting that I stay away completely. She showed me that our task on earth is to push our auras out, to be large energetically. I wanted to live in such a way that my aura expanded and I wanted to support the expansion of other people's auras. In my dealings with Larry, I felt hateful, weak, and small.

There was no way I could do what was necessary to be certified with these feelings prominent in my energy field. My rage was so great that I might be tempted to vent it and create another awful experience. I

---

[35] Mayer, p. 215.

feared that I could not keep my fury in check and would again be the cause of my own failure.

I also found myself wishing that the grapes would be sour, and that Larry would not want them anymore. I wanted him gone at all costs. This attitude weakened me further and I was not drawn to the ecstatic songs of the spirits. In fact, I was not drawn to any songs. I called B.B. and told him of my concerns.

"Probably the better path is a spiritualized one," B.B. concurred. "The preps open the plants to our thoughts. Better to create the spiritual environment so that Larry may be transformed, or not, but also so you are not hurting yourself or the earth." He offered to do a Hawaiian clearing ceremony called a *ho'oponopono*, for Donald, Larry, and me. I agreed and felt lighter in my mind and body almost immediately. Importantly, the heaviness around the Institute lifted, too.

I busied myself with finding duck manure for the compost, with turning "shit" into something nutritive. Then I dreamed that I was slaughtering cats for the purpose of making biodynamic preparations. Knowing of B.B.'s interest in dreams, I called him. Before I could state the reason for my call, he asked how I was feeling toward Larry.

"Better," I said.

"I did the *ho'oponopono*," he said, "and cleared the energy with Larry, Donald, and you. The negativity is over. You can bring it back, but it is over for now." He said that it stemmed from past lives in the Roman times, when Larry and I were merchants and Donald was a sea captain. He said that he wasn't sure who screwed whom, but Larry and I were involved.

When I told him of the cat dream, he said that cats symbolized dependence within independence. Using the cat body for the biodynamic preps was a way of composting these issues, redefining and healing the relationship with Larry. Like it or not, we were in a contract with Larry, not independent of him. B.B. also said that cats are highly intuitive, aware of everything in the environment. We should apply such vigilance in the preparations.

After hearing this, I felt calmed, contained, and grateful. On the surface, the situation seemed simple. We had signed that contract with Larry, not carefully considering the ramifications for us. Now we were paying the consequences and setting new boundaries. On another level, however, forces were at work that we needed to be aware of if we wanted relations with Larry to improve.

When I told Donald of my conversation with B.B., he asked, "Do we say anything of this to Larry?"

"You mean of the Roman times?"

"No," Donald said, "of healing, and that what we are working to do is to heal ourselves so that we can grow grapes. That we want to heal our relationship with Larry, too, toward that end."

We both knew Larry wasn't the type who would understand this, yet I liked Donald's articulation of the question. This was a healing process: the cat energy needed to be in the preps and we needed to recognize limits and attend to ourselves more in the making of business agreements. I trusted that this approach would bear fruit over time.

The dark gift of this land deal, and B.B.'s intervention, was that I felt my energy freed up to go before the Certifying Board. Yes, my previous Board meeting had been exceedingly painful, but I had done a lot of work on my part concerning that day. I had underestimated the impacts of my analytical conflict with Benjamin, of losing Don, and of the chain of events before and after these major shifts. I needed to learn other ways of dealing with rage, and to take on not only the personal dimension but the archetypal as well, which required a spiritual teacher.

Dealing with the archetypal aspects of rage is quite distinct from dealing with personal rage, and different tools are required. Don knew this, Norma knew this, B.B. knew this. Benjamin did not. I had had to sacrifice a kind of cat energy in the process of certification, too, surrendering my hope of an easy and positive outcome. I had had no idea going in that my training and certification would include such violent aspects.

There is much over which we have no control. Bad things happen and we suffer. In the process, we encounter things we need to face in order to get larger, and this is often painful. Our land dealings offered the most recent opportunity for this expansion of my spirit. My first experience with the Certifying Board had been disappointing, but it had caused me to grow.

Later I told a friend that the final Certifying Board meeting was like a long ski run downhill, fast and exhilarating. Picturing where I wanted to go, I enjoyed myself all the way there. I was certified eleven years to the day from my first interview with the admissions committee. I went to my first membership meeting eleven years to the day from when Don Sandner called me, telling me of my acceptance as a candidate. Clearing the final hurdle did not feel anticlimactic; it was, simply, the end of an ordeal.

During the same week that I was certified, we settled out of court with Larry and let the situation, and him, go. It felt wonderful to have his energy released from our ranch.

We continued with B.B. and his in-the-field-training, but the following winter, after he fell into a dark depression to which we now knew he was prone, we let him go, too. It was about this same time that Natalio and his family arrived, ushering in a new millennium on our ranch, coinciding with the one on the calendar.

* * *

One morning shortly after the terrifying events of September 11, 2001, I was walking in grief through our vineyard, worried for our earth. Out of the ground shown a pink light. When I stopped, I found a chunk of rose quartz the size of a human heart. When I held the rose quartz to my own heart, I felt such love. The spirits were singing!

I had walked the land countless times and we had worked it for years. Why was this gift being presented now? It did not take long for me to understand the message. We need to spread our heart energy on the earth; that is what will bring balance. To this day, the heart crystal rests in the arms of the St. Francis on the forest edge of the vineyard. When people visit, I tell them the story of finding the crystal. We all feel the spirits then, manifesting in waves of goose bumps.

Why do I trust this supra-normal way of approaching our ranch and our lives? Because this way of growing one's soul *through recognition of Spirit* is a healing path. I know because I feel energetically larger. And in the fourteen years that we have farmed biodynamically, our grape production has doubled, the vines are thriving, and the wine made from the native yeast is superb. In recent years, the soil structure has been such that we barely need to irrigate the grapes. And the land is more beautiful than ever, shining with vitality.

* * *

We are here on earth to become more of ourselves, to *drink of the cup of forgetfulness to remember who we are*. The recognition this brings strengthens us, growing our souls. More of our spirit moves into our bodies.

When I wake in the dark of predawn, I know this. This morning there was no wind at all. A Great Horned Owl across the meadow was hooting. From the barn came the soft tinkle of goat bells. The last quarter moon shone through the branches of the valley oak outside our bedroom window. The three year-old child who wandered the grassy field west of our country church in search of the dream pond melded into my adult body. Ash reverberated within me. The past was present, and the future, too.

For a few precious moments, I felt totally present, filled with Spirit yet completely embodied. What more from life could I ask?

Twenty-Six

January 6

Epiphany: Three Kings Preparation

Natalio arrives at 1:30 pm, just as I am beginning to stir the Three Kings Preparation. He comments, "I hope the Three Kings are light!" His unintended pun strikes me as apt.

He hopes that the Three Kings are not heavy, as he is about to carry two gallons and two cups of the preparation around the perimeter of our property, a good bit of the way quite rugged. Every fifty feet he will spray the prep outward, forming a "magic circle" of protection. This spray is called a "sacrifice" spray, which offers the energies of homeopathic gold and the resins of frankincense and myrrh at the turn of the year, an invitation to the spirits to return to their plants and animals here.

But I also see a second meaning to Natalio's words, as I hope the Three Kings Preparation brings the light of illumination to all of us on this land and to all humankind, that we may live in accordance with our inner wisdom, knowing that life on Earth is sacred.

As usual, Natalio wants to talk. I was planning on silence. The instructions state that "the person stirring be fully conscious of and focused on the purpose," and I was hoping to spend the full German stirring hour contemplating the energetic task at hand. Now Natalio informs me that he plans to stir, which feels right because he also lives here and is connected to this land. He sacrifices his good energy to keep the plants growing and thriving. He should also be part of this offering.

So I try to explain the purpose: We are making a magic circle around the property for protection against the workings of opposing forces, within and without. In a certain way, by attending to the energy bodies of all that live here, we are farming soul: if not of the plants, the animals, and of the land itself, then of ourselves.

Natalio studies me. "Is this the work of the *bruja?*"

I look up at Natalio, frowning. I take a deep breath. "No, Natalio, this is *Biodynamics.*"

Natalio points to a picture of the three rather robust kings arriving at the nativity scene, part of an article from a recent biodynamic newsletter. "The Three Kings did not bring protection," Natalio says wryly. Then he adds, joking, "And they do not look thin."

We laugh together. Again, I try to explain the theory, that this preparation is for the return of spirit to earthly life: gold symbolizing wisdom; frankincense, the cosmic ethers where the spirit resides; and myrrh, the survival of death. "We are warding off anything that could stop that," I say. "We are making a promise to the spirits of our land to take care of the land and all the plants and animals that live there."

"*Bruja!*" Natalio nods. He is not joking now, and I give up trying to convince him that we are not concocting a witch's brew.

As I stir, the heady aroma of the frankincense rises. We are in the sacred circle. Our laughter is part of it. Take nothing too seriously! While Natalio stirs, I dig potatoes. Donald loves potatoes, and these are very sweet and tender.

We take turns stirring, repeating fragments of our conversation, over and over, Natalio hoping that the three kings are thin, and then contemplating: Is this *bruja* work? Meanwhile, through stirring first one way and then the other, we throw the water into chaos so that it may receive the energies of the Magis' gifts. To these gifts, we add one of our own, laughter. Donald comes to the kitchen door to observe us. The dogs also watch for a while and then, bored, go to sleep.

At 2:30 pm, Natalio strains the preparation into the backpack sprayer and takes off. "If this is a magic circle," he says, now quite serious, "then it is important to end where I begin." I wonder if he knows the layers of meaning in his words. He walks first to the valley oak from which the spiral of stars appeared to me several years ago, then proceeds west with determination. He stops and sprays once outwards, then walks ahead 50 feet and sprays again.

I open and stretch my heart to encompass the meadow as the energy of the meadow swells to meet me. The mist from the backpack sprayer catches the sun, shimmering with rainbows, then fades into a golden arc. As Natalio disappears into the forest, I imagine the arc of light becoming a sphere that contains all the living things on this land: my family and me; the animals, birds, and insects; plants in abundant variety; and even the spirit realm, where devas and deities reside.

In my vision, we are all thriving.

I picture my teachers living and dead, and those family members who have also passed, and I give thanks for them. I feel their love. Everything is a part of the web, inextricably connected to everything else. I remember Pansy's sweat lodge prayer: *Mitakuye oyasin.*

All my relatives.

The world feels larger as the new year begins.

Acknowledgments

I have had the good fortune to have a number of exceptional mentors and friends without whom these experiences would have been, at best, very different. I am forever indebted to Thursday Night Writers who suffered through many versions of the text: Jan Beaulyn, Norma Churchill, Jimalee Gordon, Elizabeth Herron, Dianne Romain, and more recently, Leah Shelleda. Colleagues Karlyn Ward, Naomi Lowinsky, and John Steinhelber offered invaluable feedback at various stages, and Mindy Toomay's editorial skills moved the book to its final form. Karlyn Ward shared the candidacy experience with me; I will always cherish her companionship, wisdom, and love. Dyane Sherwood, also a fellow candidate, offered sanity through her excellent thinking and insight.

My gratitude extends to my several teachers and mentors whose guidance remains with me on a daily basis, although several have passed, and to "Benjamin" who birthed me into a new life, however difficult that birth was. I will always be grateful to Betty Meador who, at a dark juncture, reminded me to first believe in myself and in the guidance I was receiving from the unconscious.

Last, but not least, my enduring gratitude to my family: my husband Donald and sons, Jesse and Casey, and now their families, all who remain a grounding rod and *Pleasure on Earth* throughout a long and difficult journey.

Bibliography

Baan, Bastiaan. *Lord of the Elements: Interweaving Christianity and Nature*. Edinburgh: Floris Books, 2013.

Bly, Robert, trans. *Kabir: Ecstatic Poems*. Boston: Beacon Press, 2004.

Damery, Patricia and Naomi Ruth Lowinsky, ed. *Marked by Fire: Stories of the Jungian Way*. Carmel, CA: Fisher King Press, 2012.

Edinger, Edward F. *Ego and Archetype*. Baltimore, Maryland: Penguin Books Inc., 1974.

———. *Anatomy of the Psyche: Alchemical Symbolism in Psychotherapy*. Peru, IL: Open Court Publishing, 1985.

Hannah, Barbara. "The Beyond," *The Cat, Dog and Horse Lectures and "The Beyond."* ed. Dean L. Frantz. Wilmette, IL: Chiron Publications, 1992.

Henderson, Joseph L. *Thresholds of Initiation*. Wilmette, IL: Chiron Publications, 2005.

Jung, C.G. "Everyone Has Two Souls," *C.G. Jung Speaking: Interview and Encounters,* ed. William McGuire and R.F.C. Hull. Princeton, NJ: Bollingen Series XCVII, Princeton University Press. 1977.

———. *Memories, Dreams, Reflections*. New York: Vintage Books, 1965.

———. "Paracelsus as a Spiritual Phenomenon," *Alchemical Studies CW13*. Bollingen Series XX. Princeton, NJ: Princeton University Press, 1967.

———. *Letters, Vol. I., 1906-1950*. Selected and ed. by Gerhard Adler in collaboration with Aniela Jaffe (trans. by R.F.C. Hull) Bollingen Series XCV:1. Princeton, NJ: Princeton University Press, 1973.

Levi, Primo. *The Periodic Table*. (trans. into English from Italian by Raymond Rosenthal) Schocken Books, 1987.

Mayer, Elizabeth Lloyd. *Extraordinary Knowing: Science, Skepticism, and the Inexplicable Powers of the Human Mind*. New York: Bantam Books, 2007.

Naydler, Jeremy, ed. *Goethe on Science: An Anthology of Goethe's Scientific Writings*. Edinburgh: Floris Books. 1996.

Sandner, Donald F. *Navaho Symbols of Healing*. New York: Harcourt Brace Jovanovich, 1979.

————. "The Power of the Transference," *The San Francisco Jung Institute Library Journal*, Vol.25, No. 2, 2006. pp. 15-27.

Sandner, Donald F. and Wong, Steven H., ed. *The Sacred Heritage: The Influence of Shamanism on Analytical Psychology*. New York, London: Routledge, 1997.

Sardello, Robert, ed. *The Angels*. Dallas: The Dallas Institute Publications,1994.

————. "Introduction," *Jung and Steiner: The Birth of a New Psychology*, by Gerhard Wehr. Great Barrington, MA: Anthroposophical Press, 1990.

Spinden, Herbert Joseph, trans. "Song of the Sky Loom," *Songs of the Tewa*. Sante Fe, NM: Sunstone Press, 1933.

Watson, Lyall. *Elephantoms: Tracking the Elephant*. New York: W.W. Norton and Company, 2002.

Index

Also by Patricia Damery

Snakes

Second Edition, Leaping Goat Press (2014)

ISBN 978-0-9913098-0-1 (Paperback)

ISBN 978-0-9913098-1-8 (eBook)

Goatsong, il piccolo edition

First Edition, *il piccolo* editions, Fisher King Press (2011)

ISBN 978-1-926715-76-6 (Paperback)

Marked by Fire: Stories of the Jungian Way,

Edited by Patricia Damery

and Naomi Ruth Lowinsky

First Edition (2012)

ISBN 978-1-926715-68-1 (Paperback)

ISBN 978-1-926715-74-2 (eBook)

LEAPING GOAT PRESS

Made in the USA
Columbia, SC
15 April 2023

15390162R00109